EVALUATION AND EXPLANATION IN THE

BIOMEDICAL SCIENCES

PHILOSOPHY AND MEDICINE

Editors:

H. Tristram Engelhardt, Jr.
University of Texas Medical Branch, Galveston, Tex., U.S.A.

Stuart F. Spicker
University of Connecticut Health Center, Farmington, Conn., U.S.A.

VOLUME 1

EVALUATION AND EXPLANATION IN THE BIOMEDICAL SCIENCES

PROCEEDINGS OF THE FIRST TRANS-DISCIPLINARY SYMPOSIUM
ON PHILOSOPHY AND MEDICINE
HELD AT GALVESTON, MAY 9–11, 1974

Edited by

H. TRISTRAM ENGELHARDT, JR.

University of Texas Medical Branch, Galveston, Tex., U.S.A.

and

STUART F. SPICKER

University of Connecticut Health Center, Farmington, Conn., U.S.A.

D. REIDEL PUBLISHING COMPANY

DORDRECHT-HOLLAND / BOSTON-U.S.A.

Library of Congress Cataloging in Publication Data

Trans-disciplinary Symposium on Philosophy and Medicine, 1st, Galveston,
     1974.
Evaluation and explanation in the biomedical sciences.

Includes bibliographical references and index.
1. Medicine—Philosophy—Congresses. I. Engelhardt, Hugo Tristram,
1941–   ed. II. Spicker, Stuart F., 1937–   ed. III. Title.
[DNLM: 1. Philosophy, Medical—Congresses. W1 PH614 v. 1 /
W61 T772e 1974]
R723.T7 1974   610'.1   74–34167
ISBN-13:978-94-010-1771-8   e-ISBN-13:978-94-010-1769-5
DOI: 10.1007/978-94-010-1769-5

Published by D. Reidel Publishing Company,
P.O. Box 17, Dordrecht, Holland

Sold and distributed in the U.S.A., Canada, and Mexico
by D. Reidel Publishing Company, Inc.
306 Dartmouth Street, Boston,
Mass. 02116, U.S.A.

TABLE OF CONTENTS

SECTION V / BODY AND SELF:
PHENOMENOLOGICAL PERSPECTIVES

SECTION VI / THE ROLE OF PHILOSOPHY
IN THE BIOMEDICAL SCIENCES:
CONTRIBUTION OR INTRUSION?

INTRODUCTION

This volume inaugurates a series concerning philosophy and medicine. There are few, if any, areas of social concern so pervasive as medicine and yet as underexamined by philosophy. But the claim to precedence of the *Proceedings* of the First Trans-Disciplinary Symposium on Philosophy and Medicine must be qualified. Claims to be "first" are notorious in the history of scientific as well as humanistic investigation and the claim that the First Trans-Disciplinary Symposium on Philosophy and Medicine has no precedent is not meant to be put in bald form. The editors clearly do not maintain that philosophers and physicians have not heretofore discussed matters of mutual concern, nor that individual philosophers and physicians have never taken up problems and concepts in medicine which are themselves at the boundary or interface of these two disciplines – concepts like "matter," "disease," "psyche." Surely there have been books published on the logic and philosophy of medicine.[1] But the formalization of issues and concepts in medicine has not received, at least in this century, sustained interest by professional philosophers. Groups of philosophers have not engaged medicine in order to explicate its philosophical presuppositions and to sort out the various concepts which appear in medicine. The scope of such an effort takes the philosopher beyond problems and issues which today are subsumed under the rubric "medical ethics."

The Twenty-Eighth Annual Meeting of the American Association of the History of Medicine included a symposium, "Philosophy and Medicine," which convened in Detroit on May 13, 1955. The symposiasts were Walther Riese, a neurologist and historian of medicine at the Medical College of Virginia, Victor A. Rapport, a sociologist at Wayne State University, and Owsei Temkin, an historian of medicine at Johns Hopkins University.[2] The fact that professional philosophers were not represented reflects in part the general lack of interest by professional philosophers of this century in medicine and the general lack of interest by physicians in the possible contribution of philosophy.[3]

H. T. Engelhardt, Jr. and S. F. Spicker (eds.), Evaluation and Explanation in the Biomedical Sciences, 1–7.
All Rights Reserved. Copyright © 1975 by D. Reidel Publishing Company, Dordrecht-Holland.

In this respect the symposia which will appear in the series *Philosophy and Medicine* are meant to reflect and encourage a development of concern for the general philosophical issues in medicine, a concern which was in part stimulated by the philosophical investigation of issues in medical ethics. These symposia testify to a bilateral shift in interest on the part of philosophy and medicine, and the emergence of interest in sustained investigation of philosophical issues in medicine.

The editors are cautious about claims that there is, strictly speaking, a discipline to be known as "philosophy of medicine." Though there may be such a discipline *in statu nascendi*, this has not as yet been demonstrated to the satisfaction of philosophers as well as physicians as has, for example, the claim that "philosophy of science" or "philosophy of mind" are *bona fide* disciplines. Professor Temkin was, to some degree, aware of the fact that an argument was needed to establish "philosophy of medicine" as descriptive of a new sub-discipline within philosophy. He observed at the beginning of his paper that "Anyone wishing to speak about the philosophy of medicine will, at the outset, encounter two difficulties: the vagueness of the term and the prejudice against the subject itself."[4] On the one hand, no clear notion of the definition of the field had or has emerged. The task is still one of determining the scope of interests and methods of inquiry. On the other hand, the notion of a philosophy of medicine for some recalls bitter memories of eras in which speculation substituted for empirical methods. It reflects a suspicion of philosophy by medicine as old as the Hippocratic *Corpus* itself.[5] In fact, the collapse of speculative medicine[6] and systems of medicine[7] engendered an estrangement of medicine from philosophy's prior role in medicine[8] with the result that it was no longer generally recognized that philosophers and physicians possessed intersecting conceptual interests. We are now, though, witnessing a discovery that common grounds of interest do indeed exist.

Although the movement from philosophy *and* medicine to philosophy *of* medicine is something we may witness, it is surely not a *fait accompli*. Before such is to develop, we will have to ask what constitutes a "philosophy of medicine" if there is to be a philosophy of medicine, *sensu strictiori*. Surely the definition of W. Szumowski cited by Temkin exemplifies the vagueness of the field:

The philosophy of medicine is a science which considers medicine as a whole. It studies its position in humanity, in society, in the state and in the medical schools. It embraces at a glance the whole of the history of medicine ... [9]

In contrast, there seems to be no difficulty maintaining the existence, *bona fide*, of a philosophy of biology. In his paper of 1947, Szumowski elaborated upon his definition, noting that philosophy of medicine "reveals the more general problems of the philosophy of biology. It analyzes the methodological form of medical thought, mentioning and explaining the errors in logic which are committed in medicine...".[10] Such qualifications seem more appropriate and Temkin eventually suggests that "the philosophy of medicine should present us with a medical logic, medical ethics, and medical metaphysics."[11] These are on any account very ambitious tasks.

But the centrality of medicine, its significance for man and society, invites philosophy into medicine, even if a philosophy of medicine is not yet forthcoming. As Edmund Pellegrino has suggested, a philosophy *in* medicine must be sought even when a philosophy *of* medicine is not yet forthcoming.[12] A philosophy *in* medicine, a critical examination of the concepts and presuppositions of medicine, is needed, even if a philosophy *of* medicine is not yet at hand as an authentic and independent philosophical enterprise.

There is an urgent need to make medicine an examined profession, to subject all its presuppositions and axioms to rigorous re-examination by the elenctic method – the questioning of accepted opinion and belief, the rejection of unsupported dogmas, the demand for justification of beliefs however sacred. Medicine badly needs this illumination. Its unparalleled success as a technique has brought it to the most dangerous level of dogmatism and complacency ever. Authority in the clinical situation is being uncritically transferred to the realms of values, morals and purposes.[13]

Philosophy is needed, if for nothing else, to make plain to philosophy and to society in general the significance and the limitations of medicine. And with the temptations to hubris which medicine's contemporary abilities provide, the need for critical and reflective understanding and restraint of medicine becomes ever more critical.

Philosophy is thus offered a rich domain of conceptual issues for its examination which includes such cardinal medical concepts as health and disease, concepts which are at the same time both descriptive and normative. They, as other concepts in medicine, open a Pandora's box of philosophical questions. To understand the nature of disease and the meaning of health requires understanding what is well-being for man, the good human life. And such an understanding at once involves an examination of the entire range of human values, issues of natural teleologies, questions of ethics and aesthetics, not to mention problems concerning the status of knowledge claims that one or another state of affairs is natural or unnatural, deviant or normal. Further, the impact of medicine on society is of such a scale as to raise basic questions concerning the nature of and possibilities for human societies. Because of effective contraception, artificial insemination, possible embryo transfer, and other emerging techniques, we are forced to find new ways of talking about basic social concepts such as family, father, mother, etc. By studying and modifying the functions of organs and organ systems, medicine raises questions concerning the ways in which we can usefully talk of functions and mechanisms, of the role of human organisms and their relations to society, etc. Medicine rehearses traditional philosophical questions such as the mind-body problem and the nature of causality, not only in pressingly practical circumstances, but in contexts so closely bound to basic questions concerning human goals and expectations as to be uniquely heuristic, if not uniquely revealing.

Medicine properly invites philosophical interest in that it is a bridge between the natural sciences and the social sciences, between theoretical understandings of man and praxes which structure man and man's life. As Edmund Pellegrino has phrased it, "medicine is the most humane of the sciences... and the most scientific of humanities."[14] As such, it is an empirical science which must in its generalizations make immediate presuppositions concerning the texture of the good life. It is a view of the good life which sees well-being in a fullness of scientific and technological subtlety.

Medicine is in many ways the art and science of remaking man according to man's picture of himself, according to his hopes of what he can be, freed of the burdens of crippling disease and early death. Medicine fashions the adaptations of man to his many environments, opening new ways for man consciously to control his own processes of adaptation and response to the world. Through medicine man becomes the object of his own art, fashioning his self and the embodiment of his self. In doing so, traditional questions concerning the nature and meaning of man are made to reappear in a context of urgency and immediate practicality.

Medicine, in its modern success, has become an active philosophy of man, an implicit philosophy in need of reflective enunciation, analysis, and critique.

In this volume, and within the series as a whole, attention will be focused on basic conceptual presuppositions fundamental to medicine. The goal will be to provide an examination of these ideas of philosophical interest which are central to an understanding of the biomedical sciences and professions. Medicine as the study of man in disease and health will be used as a point of departure for a richer philosophical appreciation of man and his world. Through medicine's studies of disease, the limitations of man, and through medicine's reflections on health, the possibilities of man, a wealth of potentially useful data awaits philosophical consideration. This series will focus on this wide range of material. The goal is a mutual enrichment of both philosophy and medicine. Philosophy needs to be recalled to an original dimension of its endeavors and significance as a disciplined wondering about the condition of man. Medicine needs to be acquainted with the broad significance of its enterprise as a study of and refashioning of the human condition. Philosophy must help balance temptations to timidity and to hubris.

Evaluation and explanation in the biomedical sciences was chosen as a theme for the first volume to indicate the broad range of philosophical problems in medicine. Medicine raises issues in value theory as well as in the philosophy of science and epistemology generally. Many of its philosophical issues concern important interplays of facts and values. In any event, the philosophical questions in medicine far transcend issues of medical ethics. In fact, such questions can probably best be treated in the broader context of the more general philosophical issues of evaluation and explanation in medicine. This volume is meant as a step towards a view of that broader context. The papers include historical and philosophical explorations of the interplay of facts, theories, and values in the construction of concepts of health and disease. The nature of explanation in medicine, the role and significance of the concepts of function and mechanism, and the interplay of disease, man, and society are explored. One set of papers argues the scope, significance, and possibility of a medical ethics. Central issues such as the significance of the self and the body are examined within their context in medicine. And, the significance that philosophy and medicine have for each other is discussed out of diverse perspectives from medicine and philosophy to history and theology. The volume offers widely differing approaches – analytic, phenomenological, historical – towards the goal of demonstrating the rich range of issues which medicine offers to philosophy in particular, and to the humanities in general. In its compass, it is meant to be an invitation to all philosophers to look to medicine as an important area for future investigation, and to be an invitation to medicine to take seriously the potential future contributions of philosophy.

This series, *Philosophy and Medicine*, will sustain that invitation by yearly volumes, publishing the edited proceedings of the trans-disciplinary symposia on philosophy and medicine. The second Symposium will have the theme, "Philosophical Dimensions of the Neuro-Medical Sciences," and will be held May 15, 16, 17, 1975 at the University of Connecticut Health Center in Farmington. This and future symposia will focus on achieving a serious intellectual interchange between philosophy and medicine.

The editors wish to express their gratitude to the Council for Philosophical Studies, the Franklin J. Matchette Foundation, and the University of Texas Medical Branch for their support of the First Symposium which was held in Galveston, May 9, 10, and 11, 1974. In particular, we are in debt to Truman G. Blocker, Jr., Edward N. Brandt, Jr., Chester R. Burns, Marjorie N. Huffman, Lorraine L. Hunt, Theodore Kisiel, William H. Knisely, and Irwin C. Lieb, who contributed in various and important ways to the achievement of the Symposium. And finally, and most importantly, the existence of the *Proceedings* is due to the energies of Susan M. Engelhardt, who provided everything from inspiration and ideas, to typing and editorial labors.

H. TRISTRAM ENGELHARDT, JR.
STUART F. SPICKER

July 1974

NOTES

[1] Elisha Bartlett, *The Philosophy of Medical Sciences* (Philadelphia: Lea and Blanchard, 1844); W. Bieganski, *Medizinische Logik* (Würzburg: Cort Kabitzsch, 1929); G. Blane, *Elements of Medical Logic* (Hartford: Huntington and Hopkins, 1822); F. Osterlen, *Medical Logic*, ed. and trans. by G. Whitley (London: Sydenham Society, 1855), to name a few.

[2] Professor Riese presented "Philosophical Presuppositions of Present-Day Medicine"; Professor Rapport offered his "Alchemy, Philosophy and Medicine" and Professor Temkin titled his paper "Philosophy of Medicine" which, when revised and printed, appeared as "On the Interrelationship of the History and the Philosophy of Medicine." Abstracts of these three papers were printed in the *Bulletin of the History of Medicine* in 1956 (**30** [January-February, 1956], 44–46). The first and a revision of the last of these were later published (**30** [March-April, 1956], 163–174 and **30** [May-June, 1956], 241–251).

[3] There were, of course, a number of exceptions which would have to include physician-philosophers, such as Richard Hönigswald and Karl Jaspers.

[4] *Bulletin of the History of Medicine* **30** (May-June, 1956), 241.

[5] Hippocrates, *Ancient Medicine*, chapter 20.

[6] Guenter B. Risse, "Kant, Schelling, and the Early Search for a Philosophical 'Science' of Medicine in Germany," *Journal of the History of Medicine and Allied Sciences* **27** (April, 1972), 145–158.

[7] Guenter B. Risse, "The Quest for Certainty in Medicine: John Brown's System of Medicine in France," *Bulletin of the History of Medicine* **45** (January-February, 1971), 1–12.

[8] "We were enemies of philosophy, to be sure, but not of philosophy in general, only of the cocksure, all-knowing, self-satisfied philosophy of the forties," Rudolf Virchow, "Standpoints in Scientific Medicine" in *Disease, Life, and Man*, trans. by Lelland J. Rather (Stanford, California: Stanford University Press, 1958), p. 144. "Über die Standpunkte in der wissenschaftlichen Medicin," *Archiv für pathologische Anatomie und Psychologie und für klinische Medicin* (Berlin: G. Reimer, 1880), 3.

[9] *Bulletin of the History of Medicine* **30** (May-June, 1956), 244. See W. Szumowski, "La Philosophie de la Médecine, son histoire, son essence, sa dénomination et sa définition," *Archives Internationales d'Histoire des Sciences* **9** (1949), 1138.

[10] *Bulletin of the History of Medicine* **30** (May-June, 1956), 244.

[11] *Ibid.*, **30** (May-June, 1956), 244.

[12] Edmund Pellegrino, "Medicine and Philosophy: Some Notes on the Flirtations of Minerva and Aesculapius," Annual Oration of the Society for Health and Human Values (Delivered: November 8, 1973, Washington, D. C.), 21–25. See also Dr. Pellegrino's discussion below.

[13] *Ibid.*, 22.

[14] Edmund Pellegrino, "The Most Humane Science: Some Notes on Liberal Education in Medicine and the University," *Bulletin, Medical College of Virginia* **68** (Summer 1970), 13.

SECTION I

VALUE AND EXPLANATION: HISTORICAL ROOTS

LESTER S. KING

SOME BASIC EXPLANATIONS OF DISEASE:
AN HISTORIAN'S VIEWPOINT*

At the age of three or four a child begins to ask the question, "Why?" In
so far as we can think our way into the mind of a child, we might say that
he is trying to understand the world around him and meets with puzzles
that he cannot resolve. He is seeking an explanation. This word derives
from the Latin *planus*, which means "flat" or "smooth." Explanation
renders smooth that which formerly had been rough or uneven. Before
a person seeks an explanation he must be aware of something rough or
uneven, something troublesome in the flow of experience. The explanation, when accepted, removes that rough spot and renders the flow of
experience once more smooth.

Quite obviously, people vary in the degree to which they actively seek
explanations, or, differently phrased, people vary in their sensitivity to
rough spots. We do not need to talk about the princess and the pea, but I
would offer the example of two men who may examine a knife edge that
seems reasonably sharp. One looks at it with the naked eye and is quite
satisfied, but the other examines the knife under a microscope, to find
that the apparently smooth blade really has a jagged roughness. He
wants to sharpen it further to get out the microscopic nicks not
perceptible to the naked eye. One man is easily satisfied, the other
actively hunts for irregularities that the first one misses. To transpose the
metaphor we could say that the one has more curiosity and that he
demands more detailed and far-reaching explanation than his less
sensitive fellow. Those who are exquisitely sensitive to irregularity and
unevenness are the philosophers and scientists. They are the ones for
whom the world is full of puzzles. They are the ones who, in Browning's
words

> ... welcome each rebuff
> That turns earth's smoothness rough,

and then they try to re-establish an intellectual smoothness. To do so
they find explanations.

H. T. Engelhardt, Jr. and S. F. Spicker (eds.), Evaluation and Explanation in the Biomedical Sciences, 11–27.
All Rights Reserved. Copyright © 1975 by D. Reidel Publishing Company, Dordrecht-Holland.

Why are some persons more curious than others and why does a given answer satisfy one person but not another? These questions involve the very fabric of intellectual life, but their exploration lies beyond the scope of the present analysis. More important for us is the distinction between the philosopher and the historian. Both deal with problems of explanation but there are certain crucial differences between the two approaches, and these must be clarified if we want to avoid confusion.

In philosophy explanation is a complex process whose meaning many thinkers have tried to analyze and expound. Especially active in this regard have been the philosophers concerned with logic and the conceptual foundations of science.[1] These thinkers have sought criteria that, if fulfilled, would qualify an explanation as true or valid. They define what conditions must be satisfied for an explanation to be correct.

Historians of science have also touched upon the subtleties involved, but to a lesser degree and from a quite different viewpoint. Whereas the philosophers have sought the true meaning of explanation and the way to distinguish the true from the false, the historian tries to find out what past thinkers have offered as explanations. He is not concerned whether a given explanation is in fact true, but he tries to analyze the internal and external factors that led a particular writer to offer a particular theory as an explanation. And, in addition, the historian seeks the relation between an explanatory theory and any rival theory that succeeded it. Historians thus deal with the temporal sequence of theories and not with the intrinsic truth or error of any one of them.

The philosophers of science who study the problems of explanation have concerned themselves largely with physics and the physical sciences, expecially as these have developed since the middle of the 19th century, and have for the most part paid relatively little attention to the biological sciences. To be sure, Darwin and the post-Darwinian concepts have had a good share of attention, but the biology of an earlier period – and this includes medicine – has been largely ignored. Yet the history of medicine furnishes a rich field for the study of explanation as a process and its relation to the history of ideas.

I

Explanations can vary in complexity, generality, and degree of systematization. I want first to discuss two explanatory theories of maximum generality, so general, indeed, that either of them may serve as a basic philosophy. I identify them by the two opposing terms, "supernatural" and "natural," representing concepts of ancient lineage. The two terms imply an opposition such that either is meaningless without the other, like the terms "up" and "down," or "in" and "out." Nevertheless, the basic ideas of each can be understood without circularity.

The term "supernatural" implies a personal will that brings about a specific event. This personal will belongs to some being of great power – or at least, much greater power than that possessed by mere man. This personal being, whose will operates in a particular case, is ordinarily called a god.

I want to avoid getting embroiled in the more general category of "animism," and for the purpose of this presentation I will offer a concrete example that involves a religious formulation, namely, the mature religious system that we find in Homer. We can see how this formulation serves to explain disease.

Let us refer to the beginning of the *Iliad*. Agamemnon, leader of the Greeks, had taken captive a beautiful Theban maiden whose father, Chryses, was a priest of Apollo. The father, vainly trying to get the release of his daughter, received only rebuffs from Agamemnon. Thereupon, Chryses, hoping to rescue his daughter, prayed to Apollo to wreak vengeance on the Greeks. In answer to the prayer Apollo let fly his arrows and afflicted the Grecian host with a grievous pestilence, affecting both animals and men. The Greeks, in despair, sought counsel from their own priests. A soothsayer indicated that the cause of the pestilence was Apollo's anger at Agamemnon. After complicated bickering and bitter quarrels among the Greeks, the captive girl was returned to her father and appropriate sacrifices made to Apollo, in expiation. Chryses then prayed again to Apollo who, satisfied, then removed the pestilence.

This incident can exemplify a supernatural explanation. The crucial feature is that a personal being, much more powerful than man, can by volition change the course of phenomena. By an act of will the god can bring about events that without his express intent would not have happened.

Equally important, the will of the god can be influenced by appropriate human behavior such as prayer or sacrifice. As a corollary, we may infer that the divine agent, although vastly more powerful than man, acts from motives comparable to those of man. A further corollary points to a divine inconstancy. In a given situation the god may or may not take action, and if he takes action at one time, we cannot be sure that he will do so again a second time. Divine behavior depends only in part on external events, but as much or more on divine motives that may lie far beyond human comprehension. Some men, for example, can influence the divine will more than others. But in any case there is a lack of constancy.

The view that disease has a "natural" causation contrasts strongly with the preceding view. This contrast we find sharply indicated in the Hippocratic text, "The Sacred Disease." Hippocrates – and I use that name regardless of who actually wrote the treatise – declared that this disease, epilepsy, is not "any more divine or more sacred than other diseases, but has a natural cause." [2]

According to Hippocrates, those practitioners who alleged a divine origin to epilepsy were unable to understand the disease and were comparable to faith healers and quacks who cloaked their own ignorance under a divine attribution. These charlatans prescribed their remedies in the same way. The Hippocratic text does not reject the gods but denies that the various symptoms can be attributed to gods or that special healers, by magical rites, sacrifices and incantations, can cure the disease.

Instead, Hippocrates held that epilepsy is like all other diseases and has a natural causation. Each disease has its own nature, its own character, which is intelligible and subject to investigation. Hippocrates analyzed the physiological or, as we might say today, its pathophysiological basis – the chain of physiological events that lead to a seizure. His formulation he couched in terms of the then prevailing humoral theory which, in modern terms, was wrong in virtually every respect. But this is not important for our purposes. The important aspect was his emphasis on nature.

This term, of course, is kaleidoscopic in its meanings. For the present context I would stress certain selected aspects – that the phenomena of disease are regular and predictable, not affected by capricious volition of divine beings, but uniform, intelligible, and dependable. Phenomena can be studied, their behavior formulated into generalizations. To use terms that arose much later, we can perhaps speak of "uniformity of nature" and "natural law." In contrast, belief in the personal action of god provided no uniformity and permitted no generalizations. The investigation of nature that Hippocrates so clearly grasped was the method of science. It was one method of total explanation.

II

The conflict between these alternative methods of explanation, namely, between what is "natural" and what is "supernatural", has persisted throughout all of intellectual history. For our purposes I want to jump to the end of the 17th century. The polytheism of Homer had been replaced by Christianity whose concept of a single all-powerful God was elaborated into a complex theological system. In contrast, the classical atomism of Democritus and Epicurus had revived in the 17th century and gradually superseded the Aristotelian philosophy that had dominated scientific inquiry. In the resurgent philosophy, material particles, in motion, undergoing various combinations and recombinations, served as the basic stuff for creating explanatory theories. But the existence of the immaterial still remained dominant in its own sphere. God was immaterial, so was the soul and the mind of man, and as we shall see, so was the devil.

In the course of the 17th century what we call the mechanical philosophy served as the major explanatory theory for various aspects of experience, and firmly established the significance of "nature" and "natural" explanations. Any detailed approach would be out of place in the present exposition and I will content myself with some references to Robert Boyle. In his treatise, "A free Inquiry into the received Notion of Nature," [3] he offered a series of analyses showing the extreme complexity of the term "Nature" and the various meanings that had accrued to it. He discussed especially the relations between nature and God – between the material and the immaterial. Boyle can also serve as the spokesman for the dominant thinking of the century.

His definition, offered rather tentatively and discursively, declares nature to be "the aggregate of the bodies, that make up the world"; and that these bodies "are enabled to act upon, and fitted to suffer from one another, according to the settled laws of motion"; and furthermore the laws of motion "are prescribed by the Author of things," that is, God. [4] His discussion contains three essential features. First, that nature involves the corporeal universe, made up of matter in motion; second, that these motions proceed according to definite laws; and third, that these laws were established by God who, of course, is immaterial and quite distinct from the material universe.

Boyle made quite clear that nature in his sense is not a separate being; not an *ens* in the scholastic sense; not a demiurge or subordinate deity; not an agent or director; not a being with wisdom, skill, or power.[5] Instead, nature is compared to a machine, like the familiar clockwork mechanism, whose modes of action and behavior were established by the divine artificer. The glory and power of God were thus preserved and so too the concept of "supernatural."

Ordinary experience – and I must stress the word "ordinary" – fell into the category of nature. Nature as an explanatory theory referred events to material particles in motion and therefore was mechanistic, lawful, subject to scientific investigation. But what of "extraordinary" experience – that is, events that seem not to conform to the patterns of natural laws, events that apparently run directly counter to the ordinary predictability and regularity that govern the rest of nature? Such events might be considered miracles. The sharp distinction between God and nature provided a ready explanation. God, the all powerful creator who established the laws of nature, can suspend these laws whenever he wishes. God is above nature, truly supernatural, and his miracles are supernatural events. A miracle must be explained not by natural law but by the specific will of God that upset these laws. In the 17th century no religious person questioned the genuine existence of miracles or their supernatural explanation.

God is good, and miracles, expressing his special will, are beneficial. But many extraordinary events could occur that, far from being beneficial, might be positively harmful or degrading. Various witnesses attested such alleged events as the transformation of humans into animals and back again, the calling up of spirits from the dead, the transportation of bodies through the air, the prevision of future events, the possession of knowledge that could not come by any natural means, the misfortunes that sometimes overtake the virtuous and seem to result from a personal malevolence – these and innumerable other events all seemed to violate the known laws of nature.

The orthodox explanation attributed such events to the will of a supernatural agent, not God but the devil. God is beneficent, the devil malevolent. The devil supposedly acted through the agency of certain humans called witches and warlocks, and in the 15th, 16th, and 17th centuries there occurred that intense persecution of witches that forms such an important part of intellectual history.

Explanatory theories that involved the devil and witchcraft collided with the explanatory theories of science and "natural" causation. But the persecutions that swept over Western Europe and the New World had aroused not only the passions of man but also his critical faculties. The growth of science strengthened the hand of those who wanted to apply "natural" explanations to phenomena. Perhaps the events attributed to witches really had a natural explanation.

By the end of the 17th century the witchcraft persecutions had diminished in Europe, even if not in Massachusetts, but the theoretical implications of alleged witchcraft and the relationships to scientific inquiry still needed resolution. Such an attempt at resolution we find in a little known document, published in 1703 by one of the outstanding physicians of the century.

III

In the 17th and 18th centuries the degree of doctor of medicine ordinarily required a dissertation. Customarily the professor wrote the dissertation and the student defended it in public disputation. These dissertations were included in the *Opera Omnia* of the professor. In this instance the professor was Friedrich Hoffmann who, with Boerhaave and Stahl, dominated medicine in the early 18th century. The dissertation bears the title "De Potentia Diaboli in Corpore,"[6] and deals quite specifically with the problems of explaining particular events and deciding what might be attributed to the supernatural powers of the devil and what to the activities of nature.

Concerning the existence of the devil there was no question. Hoffman, like most intellectual leaders of the period, believed in God as beneficent and omnipotent, and also in the devil as malevolent but limited in power. What can the devil accomplish? How much can he do and how might he do it? In part, Hoffmann reasoned from definitions, in what is really a scholastic method. God is all powerful. God established nature and its

laws and only God could abrogate these laws. The devil cannot negate
the works of God. Hence the devil cannot perform miracles.

He cannot bring about events that run counter to the laws of nature.
The devil cannot make witches fly through the air, he cannot transform
bodies, he cannot bring the dead to life, he cannot create living creatures
out of inanimate objects. But if the devil cannot break the laws of nature,
he can, so to speak, *bend* them. He can use the laws of nature and direct
them for evil purposes.

The metaphysical support for this concept, and the rationale that provides its explanatory force, lie in the concept of intermediate substances.
There is latent here a form of neoplatonic doctrine, namely, a scheme of
transitions or intermediaries between purely immaterial being and the
completely material. How can the devil, for example, who is immaterial,
act on body which is material? The answer lay through an intermediate
subtle matter or spirit, that is so fine that it shares in the properties of
the immaterial, and yet not so fine that it is completely removed from
the material world.

In the great world or macrocosm this intermediate substance com-
prises the air or ether that acts as connecting link between the totally
immaterial and the completely material. The devil, who is immaterial,
can directly affect the air or ether that is almost immaterial. The air, in
turn, brings about changes in the weather and all the phenomena that
depend upon the weather (such as, for example, certain plagues and
pestilences), and these in turn affect man. When, for example, the devil
brings about disturbances in the atmosphere and thereby causes hail or
destructive storms, he is not performing miracles. He is only using nature
to accomplish evil ends. By his *will* he directs the natural forces to
produce harm. He is bending nature, not breaking natural law.

The devil acts on man in analogous fashion. Comparable to air or
ether in the great world, there is in man an extremely subtle and delicate
fluid, the animal spirits, that mediates action of the nervous system. All
sensory impressions, as well as the actions of will that produce voluntary
movement, are mediated through this subtle fluid; and the mental
activities of intellection and imagination also depend on the subtle spirit.

With this concept everything falls into a neat system. In the great world
there is the ether, in man the animal spirits. The devil, by acting on the
animal spirits, can affect virtually all mental activity and the behavior

of man. By affecting the animal spirits the devil can induce bodily motions such as convulsions, bodily states such as trances, and can distort the imagination to produce various beliefs and illusions.

We might at this point re-examine the original distinction between natural and supernatural explanations. The devil cannot bring about supernatural events in the same sense as God, but the devil can bring about what we may call preternatural events – that is, events that would not have happened without the specific malevolent will that used the laws of nature for its own purposes. The important distinction is not whether we deal with supernature or with nature, but whether we deal with volition or intent on the one hand, or impersonal "law" on the other.

In one system of explanation the concept of personality represents the critical feature. Personality carries with it the power of volition, of bringing about certain actions through the process of will. With humans this power is sharply limited, but God's will is totally unlimited. The devil, however, seems to be intermediate – his will is more powerful than man's, less powerful than God's. In the present context, "power" means the ability to control those events that comprise nature, understood in the sense as discussed above – that is, an orderly system, obeying the laws that God established and that only God can change or suspend.

A naturalistic explanation, on the other hand, holds that matter and motion are sufficient to explain events, even unusual events, and that non-human volition need not be invoked as an explanatory factor. Yet, the mechanical philosophy – a generalized term for various naturalistic explanations – tried to avoid the pitfall of materialism and atheism. The philosophers who held to mechanistic explanations were careful not to deny the reality of the immaterial – that is, the existence of God and of man's soul and mind. While the mechanical philosophy could readily apply to inanimate objects, plants, lower animals, and much of human physiology, the mind and soul of man were explicitly excepted.[7] The *mens* and *anima* of man, that have played such a dominant role in the history of thought, proved a real stumbling block in the mechanical philosophy.

Terms such as mind, thought, imagination, knowledge, will – concepts that philosophers had wrestled with without ever reaching a consensus – were sufficiently vague that they could be compatible with quite diverse

explanatory theories. Did the laws of nature apply to the mind of man?
If so, to what extent? totally or only partially? From our 20th century
viewpoint we would say that in this area the philosophers simply lacked
adequate positive knowledge to reach any reasonable conclusion. But
such a statement does not accurately reflect the contemporary viewpoint.
Philosophers and philosophically-minded physicians of the early 18th
century would certainly have admitted that there is much that they did
not know, but they nevertheless did create detailed formulations to cover
the problems they faced. We might say that no generation ever
appreciates the abyss of ignorance in which it rests, an abyss apparent
only to subsequent generations. This applies to 1974 as well as 1703.

In this paper I am interested in the way that alternative explanations
gradually get sharpened and better defined. How did medical thinkers
deal with such topics as abnormal physical and mental states? How did
they evaluate evidence? What was the status of skepticism? The definition
of nature? The concepts of divinity? We cannot take up these aspects
in a systematic fashion at this time, but they furnish a backdrop for my
presentation.

IV

Hoffman's paper attempted to harmonize various factors. Hoffmann was
a scientist. He had already published his *Fundamenta Medicinae*,[8] a text
that expounded the scientific basis of medicine. His whole doctrine he
built on the basis of the mechanical philosophy, and yet he had to take
account of other factors prevalent in the intellectual climate. Various
mental phenomena, and physical phenomena clearly related to mental
states, were calling for explanation. In regard to the problems of demonic
influence Hoffmann, like a good scientist, did not rest content with
generalities but sought specific explanations for specific events. He
tried to construct a theory that harmonized with both the mechanical
philosophy and the concept of non-human will that could control events.
Hoffman tried to harmonize conflicting views. There is no question of
a rigid *either... or* type of explanation – either purely naturalistic or
purely demonic. Rather he tried to explain facts by a synthesis of
different theories.

However, when we try to invoke the role of the devil to explain particular facts, we come up against a difficult problem that haunts all ex-

planations, namely, in any individual case, what *are* the facts? In the controversies over witchcraft some writers had accepted as true various reports of alleged remarkable events. Other writers, however, more skeptical, branded the reports as falsehoods and the alleged events as frauds and chicanery. Joseph Glanvill, staunch believer in witchcraft, had defined a witch as one "who can do or seems to do strange things, beyond the known Power or Art and ordinary Nature, by vertue of a confederacy with Evil Spirits."[9] He then added, that these "strange things" were "*really* performed, and were not all Imposture and Delusion." On the other hand, there was abundant evidence of deliberate imposture. Reginald Scot, in 1584, without denying the existence of the devil, gave many examples of proven fraud. He satirized some of the events allegedly due to witchcraft. Thus, if on the farm the cream would not turn into butter, the farmer's wife might blame a witch. But, said Scot, "...chearne as long as you list, your butter will not come; especiallie, if either the maids had eaten up the creame; or the goodwife have sold the butter before in the market."[10]

But even if many cases of alleged demonic influence were shown to be fraud and imposture, this did not disprove the power of the devil and the existence of witches. Meric Casaubon forcefully declared, "there is no truth, no nor virtue, but is attended with a counterfeit, often mistaken for the true."[11] And elsewhere he wrote, "let no man... discredit the truth or reality of any business that is controverted, because the thing is liable to abuse and imposture... For what is it, if well look'd into that is not liable to abuse and imposture?"[12] And Glanvill summed up the situation forcefully, when he declared it is not valid to infer "That because there are some Cheats and Impostures, that therefore there are no Realities."[13] From particular negative instances a universal negative cannot be logically inferred.[14]

Hoffmann did not take at face value most of the reports that alleged demonic activity, yet these reports were not necessarily fraudulent attempts at imposture. Ground existed intermediate between fact and fraud. Allegations of remarkable phenomena might represent not a deliberate falsehood but a disordered imagination, and the demonic activity might have produced this disorder of imagination. But how?

The answer involves an analysis of the imagination, to determine how much this function was purely naturalistic, consistent with the mechanical philosophy, and how much it might be subject to preternatural will. We must remember that what we now call psychology, as one of the natural sciences, was then in its infancy.

The imagination depends partly on the mind, partly on the body, that is, on the sense organs that receive external impressions. Sensation reaches the mind by intermediation of the animal spirits. Hence the devil, by acting on these spirits, can affect the sensorium and the imagination, and can induce illusions such as attendance at the witches' Sabbath, ecstasies, transmutation into animals, and the like.[15] Such a formulation made a harmonious correlation between imagination, demonic activity, empirical observations, and naturalistic explanations.

If the devil, by affecting the animal spirits, can distort the imagination, why does this happen only in some persons and not in others? There must be some difference in receptivity to the devil's influence and Hoffmann, like any scientist, tried to define these variables. Because he adhered to the main tenets of the mechanical philosophy, he, like most of the early 18th century physicians, relied heavily on the circulation of the blood to provide an explanation. Thus, he declared that persons whose blood is abundant and thick, and circulates sluggishly through the cerebral vessels, are more susceptible to the activities of the devil than are those whose blood is thin, motile, and florid. Other factors such as sex, age, diet, and climate also affect the susceptibility to demonic illusion. In France, where people drink wine and engage in intellectual pursuits, there is little talk of witchcraft, but in cold northern regions, and in areas where the diet is harsh – beans, heavy bread, pork – we have many witches and abundant "demonic illusion." Hoffmann analyzed various factors but we do not need to repeat details here. Suffice to emphasize that the devil can affect persons such as witches *only if there is some prior disposition*. The power of the devil is "bound to certain laws, to a certain disposition of the body and the blood."[16] Given this disposition the devil can then exert some dominion by acting on the animal spirits.

But clearly, we must not ascribe to the devil all the delusions of the imagination that occur in every disease or that may result from the use of drugs.[17] On the other hand, we cannot eliminate demonic influence merely because physical factors are requisite and operative. Attacks that have a physical basis are not necessarily due to that cause alone. The physical disturbances can furnish the environment, and the physical defect can offer the *occasion* for the devil to exert his power.[18] This power relates especially to inducing such conditions as convulsions, spasms and violent movements, and states where the disturbance is clearly in the animal spirits.[19] In this regard, the devil can not only increase the force of the animal spirits, as is the case in convulsions, but also diminish the flow of spirits to produce privative states such as deafness or sexual impotence.

Many authors had held that diseases ascribed to demonic origin were due entirely to natural causes. Hoffmann was fully aware of this viewpoint. He also referred to the Hippocratic text on epilepsy, the "sacred disease" deemed purely natural. It is a difficult task, said Hoffmann, to distinguish so-called natural diseases having purely physical and mechanical causes from those having "magical" origin. By this he meant diseases that arose from "higher, supernatural and moral [i.e., psychological] causes."[20]

How can we discriminate between these different categories and distinguish one from the other? To help with the differential diagnosis Hoffmann provided seven criteria that to his mind indicated a demonic causation: a sudden attack in a previously healthy man, such that suspicion of poison might arise; the use of blasphemy and obscenity; foreknowledge of the future and of secret events, especially in unlearned persons; knowledge of foreign languages that the affected person had never heard; vast physical strength that greatly surpasses the normal; the excretion or expulsion of various monstrous and heterogeneous objects, such as nails, hair, wood, flint, bones, and teeth; and finally, the failure of established remedies.[21]

v

Although more than 2500 years separated the writings of Homer from those of Friedrich Hoffmann, their attempts at explanation have much in common. Homer tried to explain an epidemic that afflicted the Greeks, Hoffman some disorders of the imagination, especially those affecting certain women known as witches. For both Homer and Hoffmann the explanation involved an act of will – Apollo in the one case, the devil in the other – and without the specific volition the particular events would not have happened. Neither Homer nor Hoffmann implied that all

diseases of whatever type resulted from some act of will. Hoffmann,
indeed, was most explicit on this score – most disorders, he emphasized,
had a purely natural causation, but certain ailments, in certain individuals, at certain times, could be attributed to demonic activity. The powers
of the devil were sharply limited. The devil could not cause any disease
whatever but only specific abnormal states. Homer did not discuss any
such specific limitations for Apollo but seems to imply that the god might
have caused any disease he wished.

Although an act of will serves as "ultimate" explanation, there were
important intermediate steps. Apollo implemented his will through his
arrows while the devil's will involved a far more complicated process,
namely, particles in motion obeying certain mechanical laws, then
especially fine particles, called "animal spirits," and finally an immaterial
entity called the mind, all in a complex but nevertheless direct chain of
connection.

Homer, in attributing disease to the arrows of Apollo, did not offer
any details that today would serve as "pathogenesis." Did Apollo's
arrows, *by themselves*, convey the disease? Were they the specific
disease-causing agent in and for themselves, an agent that Apollo
merely directed? Would these arrows have caused the disease if someone
else had directed them – if, say, Mercury had stolen them and then shot
them in jest?

If we answer these questions affirmatively, then we approach a naturalistic explanation. The alternative holds that the arrows were simply an
ad hoc agency of Apollo's will which the god could have implemented in
many other ways. For example, he might have scattered handfuls of
pebbles or sprinkled drops of water, and conveyed the pestilence just as
well. Did the potency reside in the object, so that arrows served instead
of the bacteria or viruses of a much later era; or did potency reside
entirely in the personal will of the preternatural being – in this case,
Apollo? Homer did not discuss this point specifically but the general
tenor suggests that the arrows did not of themselves exert power. All
depended on the personal will of Apollo.[22]

With Hoffmann, on the other hand, the pathogenetic mechanisms
were sketched out through a chain of events involving natural forces.
Sometimes demonic will initiated the sequence that led to disease, but
any other event that could have started the causal chain would have

produced a similar result. Drugs, for example, might induce similar delusions but on a purely natural basis.

Even though totally incorrect, Hoffmann's schema indicates the formal character of an explanation. Homer's account of the epidemic (and epizootic), wrong as it is, also represents a formally valid explanation and is, I maintain, similar to Hoffmann's. To maintain this position I would offer a brief exposition of the nature of explanation.

Let me add a few additional examples of causal assertions. Thus, (1) Why did Jane lose her temper? *Because* she was upset at failing the test in school. (2) Why did John catch cold? *Because* he went out in the rain without his rubbers. (3) Why is foxglove good for certain kinds of "dropsy"? *Because* it "strengthens" the action of the heart. (4) Why did Job suffer so many misfortunes? *Because* God permitted Satan to afflict him.

In all of these we start with some phenomenon – the *explicandum* – Job's boils, Jane's temper tantrums, Uncle Henry's dropsy, John's coryza, the Homeric pestilence, the trance or convulsions of the witch.

This explicandum I might call "Area 1." When we want to explain it, we enlarge it by adding on a quite different area or, as the logicians call it, a different universe of discourse. This I designate the *explication* and will call it "Area 2." Whether the explication be "factual" or "conceptual" is not important. Area 2 may be simple or complex, factual or conceptual, but the important feature is, that Area 1 and Area 2 should join in a smooth and acceptable fashion. Area 1 and Area 2 enter into an acceptable relationship that satisfies the curiosity of the questioner. If this takes place, Area 1 is explicated by Area 2, and the conjunction of the two areas *is* the explanation. In metaphorically spatial terms we say that the conjunction makes things smooth; the questioner can pass from Area 1 to Area 2 and back with ease and satisfaction. If he cannot do so, then no explanation exists.

There is no requirement that the explication should be "correct," for judgments regarding correctness are relative to historical contexts. It is necessary, however, that there be a smooth transition, and smoothness is a subjective factor.

The historian has the task of mapping out the various areas that I designate as 1 and 2, and analyzing why at certain times men could find the passage from one to the other quite smooth, while at a later time what once seemed smooth has become mountainous and impassable, and is no longer satisfactory as an explanation.

Since all explanations are not equal, we must introduce *value* and distinguish good explanations from bad. And the historian, realizing that evaluation changes with time, must explain – I use the word advisedly – why what seems good at one time seems bad at another.

Obviously this theme has extensive ramifications but at the moment there is no opportunity to follow out the many open pathways. I expect to develop these further in subsequent studies. For the present I must rest with showing some relations of medical history to the problems of explanation.

Chicago, Illinois

NOTES

* This study was supported by a grant from the Public Health Service Research Grant LM 01804-01.

[1] There is an enormous literature on the philosophical problems of explanation. A few of the major works that have been helpful include C. G. Hempel, *Aspects of Scientific Explanation and Other Essays in the Philosophy of Science* (Glencoe: Free Press, 1965); Georg Henrik von Wright, *Explanation and Understanding* (Ithaca, N. Y.: Cornell University Press, 1971); Ernest Nagel, *The Structure of Science: Problems in the Logic of Scientific Explanation* (London: Routledge and Kegan Paul, 1971); Karl R. Popper, *The Logic of Scientific Discovery* (London: Hutchinson, 1959); Mary B. Hesse, *Models and Analogies in Science* (Notre Dame, Ind.: University of Notre Dame Press, 1966).

[2] *Hippocrates*, with an English translation by W. H. S. Jones, Loeb Library edition, 4 vols. (London: William Heinemann, 1923), II, 139.

[3] *The Works of the Honourable Robert Boyle*, ed. by Thomas Birch, 6 vols. [Facsimile edition, London, W. Johnston et al., 1772] (Hildesheim: Georg Olms, 1966), V, 158.

[4] Boyle, V, 177.

[5] *Ibid.*, pp. 162–165 and *passim*, 175, 191.

[6] The original dissertation of 1703 bears, on the title page, "Praeside Friderico Hoffmann ... submittit Godofredus Bueching, Halle, Gruner." In Hoffmann's collected works (*Opera Omnia*, in 6 vols. with 2 supplements [11 vols. in all], Geneva: De Tournes, 1741–1750), the essay appears in vol. V, pp. 94–103, the title is slightly altered to *De Diaboli Potentia in Corpora* and Bueching's name does not appear. The text shows some polishing and stylistic improvements, compared with the original edition. I have taken the quotations from the edition appearing in the *Opera Omnia*.

[7] Boyle, V, 166.

[8] Friedrich Hoffmann, *Fundamenta Medicinae* [1695], trans. by Lester S. King (London: Macdonald, and New York: American Elsevier, 1971).

[9] Joseph Glanvill, *Saducismus Triumphatus, or Full and Plain Evidence Concerning Witches and Apparitions* [1689] (Gainesville, Florida: Scholars' Facsimiles and Reprints, 1966), p. 269.

[10] Reginald Scot, *The Discoverie of Witchcraft* [1584] (1930; reprint ed., New York: Dover Publications, 1972), p. 6.

[11] Meric Casaubon, *Of Credulity and Incredulity in Things Natural, Civil, and Divine* (London: T. Garthwait, 1668), p. 31.

[12] *Ibid.*, pp. 164–5.

[13] Glanvill, p. 87.

[14] Before we apply an explanatory theory we should be quite clear just what we are trying to explain. To use current jargon, in any given instance what are the "facts" of the case? "Facts" – I must enclose the word in quotation marks – come in bundles and clusters. For a given cluster we may find a satisfactory explanation. But if to that cluster of events we add a few additional facts, we may produce thereby a totally different situation, so that the explanation satisfactory before the addition might not suffice for the newly constituted bundle. The data relevant to witchcraft furnished many obvious examples. To use the illustration that Scot offered almost four centuries ago, if the dairy maid cannot make cream turn into butter, perhaps a malevolent spirit had influenced the result. But if we add the additional fact that the girl had drunk most of the cream before starting to churn, then we have a new situation and we will find some alternative explanation more satisfactory. The problems of explanation, in large part, depend on what cluster of events we try to explain.

[15] Hoffmann, *De Diaboli Potentia...*, §18.

[16] *Ibid.*, §19.

[17] *Ibid.*, §19.

[18] §20.

[19] §21.

[20] §24.

[21] §24.

[22] In this connection see the discussion in my *The Growth of Medical Thought* (Chicago: University of Chicago Press, 1963), pp. 13–17.

CHESTER R. BURNS

DISEASES VERSUS HEALTHS: SOME LEGACIES
IN THE PHILOSOPHIES OF MODERN
MEDICAL SCIENCE

I. INTRODUCTION

Through centuries of Western civilization, medical and non-medical savants have asked numerous questions about the nature of health and disease in human beings.

Ancient Babylonians and Egyptians ascribed a substantive existence to disease. Human diseases represented the activities of supernatural beings or were caused by these beings. *The Exorcist* demonstrates the persistence of this concept through centuries of Western culture.

Yet, do diseases really exist as entities, like demons or frogs or leaves or rocks? If so, what are their characteristics and how do they differ from their opposites – healths? If they are ontologically real, do diseases have particular histories? If diseases are substantive beings, what are their causes?

As can be learned best from reading *On the Sacred Disease*, Hippocratic physicians transposed the problem of disease etiology from a supernatural realm to a natural one. The causes of disease resided within the natural processes of each individual human organism. Sickness was always a problem of an individual human regardless of the ultimate cause of the human's existence or disability.

For Greek physicians, the natural causes could be external or internal. It was difficult for them to believe that the pestilential diseases that decimated hundreds did not originate from a source external to themselves. On the other hand, they observed diseases that seemed to be caused by processes occurring within the individual, such as the formation of an abscess. Regardless, the individual human was the ontologically real being for most Greek physicians. The perceived functions and activities of a given individual changed. Some of these changes were called diseases.

Did this approach mean that health, the opposite of disease, was merely the absence of the changes or events called diseases? Or was health a timeless, changeless ideal?

H. T. Engelhardt, Jr. and S. F. Spicker (eds.), Evaluation and Explanation in the Biomedical Sciences, 29–47.
All Rights Reserved. Copyright © 1975 by D. Reidel Publishing Company, Dordrecht-Holland.

The idealization of clinical observations has recurred throughout Western medicine. Physicians have been chastised repeatedly for confusing nominal reality with physical reality; for committing certain ontological fallacies. The ones most frequently mentioned involve the transmutation of intellectual constructs into perceptual realities, such as the humors of ancient Greece.

It is inappropriate – so the argument goes – to allow such imaginative ideas to become confused with the perceptible realities of disease. "Disease" and "health" are nothing but the products of human minds. They are not ontologically real. They are simply ideological constructs, convenient for argument and action, but signifying only a nominal reality.

The ontological status of health and disease with appropriate space-time components; etiological considerations of disease and health whether viewed as ontologically or nominally real; the view of health as the logical opposite of disease – these are fundamental legacies in the philosophies of Western medical science. This re-examination of these legacies will focus on some answers given to the aforementioned questions by medical thinkers in the modern era. These answers have been offered within four contexts: Organismic Dis-Ease; Organismic Ease; Mental Dis-Ease; and Mental Ease.

II. ORGANISMIC DIS-EASE

An individual human has or is a body that is composed of differing parts. Medical practitioners, ordinarily, have not denied the ontological significance of this assertion.

Some ancient Egyptians organized their observations about human wounds in anatomical terms, beginning with the top of the head and proceeding toward the toe. Since their time, most somatic conceptualizations of human diseases have been created with anatomical frameworks.[1] For fifteen hundred years, medical practitioners believed, erroneously, that Galen had described the parts of the human body. Andreas Vesalius, a professor of surgery, realized that the anatomical truths ascribed to Galen were not based on dissections of human bodies. After the publication of his book in 1543, the anatomical idea gradually permeated most somatic conceptions of human disease.

A multitude of observations emerged from countless necropsies, dissections, and operations performed during the three centuries after Vesalius. Physicians located disease in a bodily part, described the changes which had occurred in that part, classified types of diseases in anatomical terms, and attempted to explain diseases with anatomical language. But anatomical analysis was not sufficient, as Galen had realized many centuries earlier.

Galen had designated the disorders of functions of a bodily part as its specific symptoms. To a particular individual, these symptoms represented functional changes, such as muscular weakness or vomiting. In observing a patient, the physician might perceive other changes in function or structure that were designated as "signs" of disease, such as a lump on the neck or a discoloration of the skin. Since different organs were involved, functional disturbances in different parts of the human body produced different clusters of "symptoms" and "signs." The study of these symptoms and signs, semeiology, has been an essential part of medical science since the days of Hippocrates and Galen.

If symptoms and signs are evidence of abnormal functions, would it not be possible to classify diseases in terms of these symptoms and signs?

In the 17th century, Sydenham believed that it was possible to identify a "natural history" of a given disease, that is, a strictly objective description of the naturally spontaneous sequence of abnormal functions represented as symptoms and signs. He believed that every disease could be characterized with its specific history. A score or more of 18th century clinicians, including Sauvages, Linné, Cullen, and Pinel, created elaborate and detailed catalogues of diseases according to symptoms and signs. Some of these individuals allowed their labels to become the diseases; that is, they committed the ontological fallacy of reification. Fever was not just a symptom or sign, but an essential disease. It should not be forgotten, however, that most of these individuals sought a functional or physiological classification of diseases that would permit effective diagnosis, therapy, and prognosis. Even so, the latter goals would not be reached until a marriage had occurred between anatomical and physiological analyses.

By correlating the structural changes in the organs of a dead individual with the signs and symptoms that had been observed when that person was alive, a few clinicians of the early modern era attempted to integrate anatomical and physiological considerations into clinically useful somatic concepts of human disease. These attempts culminated with the work of Morgagni published in 1761. Morgagni not only juxtaposed descriptions of clinical symptoms with necropsy observations, but he also attempted to explain the clinical symptoms with anatomical referents.[2]

Recognition of the importance of this kind of correlation stimulated an enormous amount of clinical activity that has persisted to this day. Championing careful observation of sick persons, clinicians discovered new investigative techniques such as percussion of the thorax and stethoscopy. Others began to concentrate their diagnostic efforts on particular organs or organ systems. Others, dissatisfied with the difficulties of correlating symptoms and signs with changes in dead organs, selected certain anatomical or physiological or chemical features of human organs as the basis for speculative explanations of somatic diseases. Some, like Hoffmann, Stahl, and John Brown, committed reificational and other ontological fallacies in their quests for scientific certainty and clinical security.[3]

By the middle of the 19th century, medical scientists and practitioners had inherited centuries of observations and reflections about the somatic nature of human disease. These efforts had been motivated by the assumption that human diseases would be explained by studying the structures and functions of the various component parts of individual human bodies. An enormous amount of empirical knowledge had been acquired and clinicians had become more expert in diagnosis and treatment, particularly surgical.

By 1850, though, the emphasis on anatomical observation and explanation began to be replaced by the more aggressive viewpoints of experimental physiologists who desired to learn more about the nature of physiological processes by manipulating the structures and functions of animal and human organisms. Many emphasized that diseases were simply the deranged physiological processes of an individual human. Furthermore, they argued that rational physicians would have to reject ontologically fallacious notions of diseases and focus on "diseases" as functional alterations in the life processes of individual humans. Doctrines of specific diseases did not belong in this framework.

Ironically, however, the application of physiological techniques to problems of human functions enabled some clinicians to view abnormalities of these functions as specific diseases. Alterations in function of specific organs could occur without visible changes in structure. These functional alterations could be postulated as the causes of symptoms experienced and signs observed. Therefore, they were ontologically real disease entities. For example, gastric insufficiency, either in terms of motor weakness or chemical disturbance, was as much a disease entity as gastric ulcer. Clinical physiologists of the late 19th century were no more exempt from committing ontological fallacies than their anatomical or nosological predecessors had been.[4]

This clinical inclination toward ontological confoundedness was reinforced with the rise of bacteriology in the last three decades of the 19th century. The germ theory of disease diverted attention away from the life processes of individuals and toward diseases as entities. For example, it became as, if not more important to destroy the pneumococcus, than to restore pulmonary functions in a given human. In fact, it was believed that physiological processes could not be restored in any human organ without destroying the invading and reproducing microorganisms. On the other hand, it would be possible to prevent the occurrence of these human diseases by excluding the microbes from the human environment and by utilizing the same methods of immunization that had been employed with smallpox for decades. Again, extraordinary consequences followed from the bacteriological discoveries. Perceptually, bacteria were real. As causes for numerous human diseases they were as tangible as demons were intangible. With this new knowledge, somatic concepts of diseases were rapidly re-fashioned, especially by pathologists.

Pathologists emerged during the 19th century as the foremost students of disease.[5] The "principles" of this science, "pathology", represented generalizations gleaned from anatomical, physiological, and bacteriological investigations of the parts of individual animal and human bodies. The ideas of pathologists reinforced the notion of diseases as entities that could be explained by an analysis of the various bodily parts of given individual humans. Pathologists, like anatomists, physiologists, biochemists, and bacteriologists, did much of their experimental work with non-human animals. Accordingly, their concepts of disease remained biological in orientation, and represented a self-imposed segregation from the psychological and social variables of being human. The task of exploring organismic dis-eases was challenge sufficient, and most clinicians agreed.

Clinicians of the 19th and 20th centuries approached their sick patients with the anatomical, physiological, and pathological data afforded by the basic scientists. Anatomical and etiological criteria were utilized in the preparation of schemes for classifying diseases, that is, in preparing nosologies and standard nomenclatures.[6] During the past 100 years, the anatomical subdivisions in these various classifications did not change, with the exception of a new category called "Psycho-Biologic Unit." Because of new discoveries in bacteriology, physiology, and biochemistry, changes occurred in etiological categories. These changes, though, did not alter the fundamental assumptions of disease ontology and etiology. Nor did they prevent scientists and clinicians from continuing to commit ontological fallacies as they investigated organismic dis-eases.

Another set of fallacies, more logical than ontological, was apparent in investigations of organismic ease.

III. ORGANISMIC EASE

Usually, Organismic Ease or Health is viewed as the conceptual opposite of Organismic Dis-Ease. This view pervades centuries of clinical medicine in the West.

According to Galen, health was the absence of pain and the absence of obstacles that hindered daily activities. Health was identified with "natural" or "according to nature" and connoted a symmetry of the components and processes within the body of an individual human organism. Health, the Ease resulting from this symmetry, was preserved by a carefully determined regimen of eating, drinking, sleeping, exercise, sexual activities, and emotional expression.[7]

In the medieval era, the health of the body and the health of the soul were separated. The body could be healthy; the soul sick. Or both could be experiencing the painful consequences of sin.

In the early modern era, the idea of health was naturalized and somaticized. Some diseases continued to be viewed as the results of immoral activities. But the idea of health as the consequence of goodness was replaced with the idea of health as the consequence of knowledge. Men became ill because they did not understand their bodily processes. To provide this understanding, physicians, teachers, and government authorities became health educators during the Enlightenment. Natural perfectibility was possible for every man who, understanding the functions of his body, sought to maintain and enhance them. Ease was attainable.[8]

An organized response to this ideal of health was made at Boston in 1837. The American Physiological Society was established by several lay persons who wished to learn "that part of Human Physiology which teaches the influence of air, cleanliness, exercise, sleep, food, drink, medicine, etc., on human health and longevity."[9] Ordinary, but enlightened, American citizens wanted to understand human physiology and appropriate hygienic practices.

Traditionally, though, physiology was an integral part of medical science. Orthodox physicians knew more about human physiology than any other group, presumably. Normal physiology was necessary for understanding symptoms and signs as indicators of abnormal physiological processes. Normal physiology was also necessary for understanding hygienic rules and practices as "applied physiology." From the days of antiquity until the latter part of the 19th century, medical students had been expected to understand hygiene as "applied physiology" and, as practitioners, to utilize this understanding in promoting the health of the individual. Without question, cures of the individual's diseases were efforts toward the promotion of his health. However, such efforts were not sufficient, if the principles of hygiene were to be applied properly by the practitioner.

It is no accident, therefore, that books on hygiene were included in the "physiology" section of the report presented in 1848 by the Committee on Medical Literature of the American Medical Association.[10] During the first half of the 19th century, textbooks of physiology frequently mentioned the relationships of physiology and hygiene. "The art which most directly springs out of the science of Physiology is that of Hygiene which may be defined as a system of rules for the preservation of the body and health, deduced from the principles by which its actions are governed."[11] Physiology was to hygiene as pathology was to therapeutics. But few scientists or clinicians were able to correlate either pair, and hygiene included considerably more than "applied physiology" for the few American practitioners who cared at all.

For most of these men, such as Elisha Bartlett, John Bell, and Robley Dunglison, health and morality were intertwined as intimately as they had been in the medieval era. "Rules for the preservation of beauty" were the "same rules to be followed for the support of health," both physical and mental.[12] These were also the same rules required by each individual "to maintain his ethical and religious relations with his fellow men."[13] If these assertions were true, practitioners who professed to be hygienists were required to understand the "physiology" of the mind and the soul. This philosophy of health and hygiene also justified invasion of the "territorial imperatives" of the professional caretakers of man's mind and soul, in particular, clergymen. How could this philosophy be synchronized with the hallowed tradition of the physician as the caretaker of man's body? For many American clinicians of the 19th century, there was no synchrony and the traditional ideals of health and hygiene became peripheral, if not irrelevant to the practice of medicine.

As previously documented, study of diseases was the fashion of the day. Diseases were to be identified anatomically, physiologically and etiologically. This framework could be used for classifying diseases, and classification facilitated diagnosis. Correct diagnosis increased the probability of therapeutic cure. By accurately identifying and adequately curing particular somatic diseases, physicians believed that they did all that was necessary in promoting human health. Health or Organismic Ease was the result of removing Organismic Dis-Eases.

This view of health as the opposite of disease received an extraordinary impetus from the germ theory of disease. During the last three decades of the 19th century, bacteriological discoveries offered a new rational basis for specific hygienic practices, both personal and public. Hygiene, as applied physiology, was transformed into preventive medicine, as the fight to preserve health by destroying the microbes that produced contagious diseases. In addition, personal hygiene was subsumed by public hygiene.[14]

In schools of medicine, hygiene was taught primarily as a matter of infectious disease control. For example, lectures in hygiene were given in the physiology department at the University of Texas Medical Branch in the 1890's and even during the first decade of the 20th century. But during the second decade of this century, these lectures were transferred to a Department of Preventive Medicine, a department that later combined with a Department of Bacteriology, again illustrating the fundamental impact of bacteriology on hygiene and the transformation of hygiene into preventive medicine.[15]

As lectures in hygiene disappeared from physiology departments and as a new breed of experimental physiologists began to occupy teaching positions in American medical schools, the view of hygiene as "applied physiology" disappeared altogether. Physiologists seldom presented their data as a comprehensive view of the healthy processes of an individual human organism. "Normality" was assumed and "hygiene" ignored. Physiological facts – based primarily on animal and not human experimentation – were presented randomly and future physicians selected from them either to pass exams or to apply them to particular clinical problems.[16]

Professors of preventive medicine or public health offered the only viewpoints that might have been described as health oriented. Yet, in the early decades of this century, their views of "health" emphasized the prevention of infectious diseases in groups (epidemiology). Few gave any systematic attention to the study of a healthy individual human and to personal hygiene as such; certainly not as "applied physiology" in the way that hygiene had been understood by medical thinkers prior to 1870, or as a set of values that could be used in assessing human behavior as a whole.

Since 1950, a more comprehensive concept of health has been resurrected by some educators and clinicians who are concerned with broad programs of health care; programs characterized in at least three ways: (1) care of individual patients either in comprehensive or holistic terms, (2) total care of families via family medicine specialists, and (3) overall care of communities with a variety of health care professions.[17] In most of these programs, though, health maintenance is still viewed as the prevention of organismic dis-eases and not the establishment of life styles that produce states of "complete physical, mental, and social well-being." But the evaluative nature of the concept of health is omnipresent, as denoted by this definition of health given in 1946 by the World Health Organization.

To acquire further insight into the ontological and logical paradoxes of Western approaches to health and disease, medical studies of mental disease and mental ease will be examined now.

IV. MENTAL DIS-EASE

The Hippocratic transition from a supernatural to a natural etiology redirected attention to man himself as the arbiter of his destiny. Hence, disease could be produced within an individual by the adaptive conflicts of his soul, mind, or psyche.

That the mind could be diseased or could be a cause of somatic disease was explored by Galen. He had postulated the existence of three regulatory entities in an individual: the natural, vital, and rational souls. The passions and judgments of the vital and rational souls could become deranged. Greed and envy were abnormal passions. Ignorance about the goals of life led to false judgments. Knowledge, self-evaluation, and self-regulation were methods for curing these passions and errors, according to Galen.[18]

A small number of mental diseases was described by other authors in antiquity. Melancholia and mania were recurring entities. These were often listed as diseases of the head or with diseases of the head or brain. Not all of the entities were somaticized. Paul of Aegina and St. Thomas Aquinas recognized demonomania or behavioral aberrations caused by the actions of demons.[19]

Supernatural concepts were utilized in medieval approaches to health and disease. In accounting for behavioral peculiarities, such religious concepts were fundamental. Not infrequently, a mentally ill person was thought to be evil or possessed. Even those physicians of the Renaissance who continued to identify mental diseases with the brain, such as Jean Fernel, or who offered innovative classifications of behavioral abnormalities, such as Paracelsus or Felix Platter, did not deny the existence of demoniacal possession. In these latter classifications, though, we begin to get some glimpses of the ontological reality of the mind divested of theological considerations.

In the 17th century, Descartes reinforced a conceptual separation of mind and body. Ontologically, the mind was separated from the body by Descartes and other philosophers who wished to investigate the faculties of the mind as natural entities.

Some physicians of the 17th and 18th centuries responded to the claim that mental functions of an individual human deserved observations and explanations that were not dependent on parts of the brain or sins of the soul. This is illustrated strikingly by the medical nosology of Sauvages (1706–1767). Sauvages viewed the majority of mental dis-eases as disturbances of extra-cerebral origin, or disturbances of instinctual and emotional life, or disturbances of intellectual life.[20]

Among various philosophers of the period, including Kant, non-physical and non-theological approaches led to diagnostic labels and classifications of mental diseases as aberrations of the faculties of the mind or soul. Far into the 19th century, some philosophers of the mind considered their studies incomplete without speculative investigations of "imperfect and disordered mental action."[21] The ratiocinations of these philosophical views of mental diseases profoundly affected physician-students of the mind, such as Pinel and Rush.

The nosological simplifications of Pinel and Rush – although curtailing some of the abuses of nosologizing – were excessively rational in nature. Their efforts, though, did not check the quest of clinicians for a plurality of mental diseases nor the quest of medical scientists for explanations of these diseases in anatomical and physiological terms.

At the turn of the 19th century, Cullen, Pinel, Rush and others began to view mental diseases as functional disorders of the nervous system. Cullen's nosology is a formidable example of an attempt to classify mental diseases in terms of symptoms and signs representing derangements of the nervous system, particularly as they are manifested in disorders of sensation and motion. This anatomico-physiological approach was carried to elaborate extremes in the 19th century, as mental diseases became identified exclusively with anatomical or functional derangements of the brain.[22]

This somaticization of mental diseases was counterbalanced by a shift in clinical approaches that occurred toward the latter part of the 19th century and in the early years of the 20th century. Many clinical textbooks of that period utilized dual terms: nervous *and* mental diseases.[23] This shift is also illustrated by examining the American nomenclatures of diseases adopted during the past 100 years.[24]

In these classifications, a major change occurred in the category called diseases of the nervous system. In the nomenclature published by the American Medical Association in 1872, only a few types of mental illness were listed in this category. This number increased significantly in the AMA's *Standard Nomenclature of Disease* published in 1933. In the following year, the American Psychiatric Association adopted a new classification that was incorporated into subsequent editions of the AMA's *Standard Nomenclature*.

Neural, organic, and somatic frameworks structured fifteen of the categories identified by the American Psychiatric Association. A more strictly psychological model underlay the remaining categories that included psychoneuroses, manic-depressive psychoses, schizophrenia, paranoid conditions, other psychoses, and primary behavior disorders. This dualism was continued in the *Diagnostic and Statistical Manual of Mental Disorder* published by the American Psychiatric Association in 1952. One part is a list of "disorders caused by or associated with the impairment of brain tissue function," and the other is a list of "disorders of psychogenic origin or without clearly defined physical cause or structural change in the brain."

Neuroses *and* psychoses are categories based on the ontological realities of mental diseases, as nervous system derangements on the one hand, and as psychological derangements on the other. Clinically, psychiatrists acknowledge a realm of mental dis-eases that is causally related to organismic dis-eases. On the other hand, they recognize a realm of mental dis-eases that are ontologically and etiologically distinct from organismic dis-eases. The strict somaticist would argue that this latter realm is actually the finest group of ontological fallacies ever created by medical thinkers.[25]

V. MENTAL EASE

Recall the broad definition of hygiene that had been used by some American physicians in the first part of the 19th century. Hygiene denoted a comprehensiveness of viewpoint that included man's body, mind, and soul. However, no physician actually attempted to transform this definition into practical judgments that could be used in caring for man's body, mind, and soul. These early American physicians discussed diet, beverages, climate, seasons, and other factors that pertained primarily to the hygienic care of the human body.

A new viewpoint emerged as the number of patients in state insane asylums grew steadily and as a growing number of American physicians accepted responsibility for their care. The incidence of mental illness had to be lessened by greater attention to mental hygiene.

In 1863, Isaac Ray, one of the more outstanding of these physicians, defined mental hygiene as "the art of preserving the health of the mind against all the incidents and influences calculated to deteriorate its qualities, impair its energies or derange its movements."[26] In abundant detail, Ray discussed the cerebral, physical, mental, and social influences that undermine mental health. In keeping with the earlier definitions of hygiene, he emphasized the mental and moral aspects of being human. Among numerous exhortations, for example, he encouraged parents to provide suitable moral values for their children. This would be the best method for preserving their mental health.

A hasty glance would suggest that these ideals were rejected by the alienists and neurologists of the 1880's and 1890's. As many physicians of this period attempted to reduce all mental diseases to dysfunctions of the brain, they also attempted to make mental and physical hygiene identical. Only in theory, however, since most physicians acknowledged the role of "moral" or "psychological" factors as proximate causes of mental dis-eases and since most doctors recognized the importance of moral and psychological measures in therapeutic efforts.[27]

Furthermore, as with physical disease, insanity was viewed as a disease of ignorance. A knowledge of the laws of the human mind would enable an individual to preserve his sanity as well as to prevent symptoms of insanity. Mental hygiene was "applied psychology" just as bodily hygiene was "applied physiology."

Even the widespread belief in hereditary predisposition to insanity did not undermine the importance of moral prophylaxis in the philosophies of the physicians. Mental health was determined by the environments of the home, the school, and the church. Obedience, self-control, self-reliance, and devotion to duty were the mental qualities necessary for preventing insanity. These and other moralistic views were utilized by mental hygienists in urging reforms of American schools and other American institutions.

Attributes of mental health were almost identical with the standards of conduct that undergirded American society as a whole. The mentally healthy were conforming, obedient, moral, and industrious citizens – like those who had founded the country some 250 years previously. In the latter part of the 19th century, "mental hygiene" or "mental health" was the rubric for transmitting the religious and political values of old.

After the turn of this century, there were few, if any, attempts to make mental hygiene identical with physical hygiene. Mental hygiene was buoyed by the rapidly developing science of psychology. Mental hygiene also became a part of the transformation of personal hygiene into public hygiene.

Mental hygiene was that portion of public health concerned with all aspects of mental illness. Epidemiological surveys, public education, improved hospital facilities, and scientific research were connoted by this term. For example, mental and social hygiene were identified as important areas of study in the original description of the Johns Hopkins School of Hygiene and Public Health founded in June of 1916. In 1934, these studies became more significant as a statistical survey of the incidence of mental diseases in the Eastern Health District of Baltimore was undertaken.[28] Note, though, that mental hygiene was achieving scientific respectability by determinations of the incidence of particular mental diseases and not by investigations of the "normal physiology" of an individual mind or person. In becoming a part of public health, the personal dimension of mental hygiene was being transformed into clinical and societal indices in the same way that bodily hygiene had been transformed into the laboratory reports and statistical curves of preventive medicine.

The science of psychology exerted equal, if not greater, influence on mental hygienists. Via the work of Adolf Meyer, especially, psychological viewpoints were introduced into the philosophies of clinicians caring for the mentally ill. These same men also introduced mental and instinctive factors into their ideas of mental hygiene. In *The Principles of Mental Hygiene* published in 1917, William Alanson White offered three introductory chapters that dealt with "mental mechanisms" and other psychological concepts. For White and others, mental hygiene had become "applied psychology" as well as "preventive psychiatry." It would be necessary to understand the basic mental functions of an individual human in order to "intellectually approach the various practical applications of the principles of mental hygiene."[29] Thereby, insanity, criminality, prostitution, vagrancy, and a host of other "sick" psychological and social conditions could be prevented.

Promoting mental health, as well as preventing mental illness, became a fundamental objective during the Progressive Era. A sound education was still viewed as the most important way to accomplish both objectives. New ideas about the nature of childhood and new methods of pedagogy did not alter this fundamental belief. As Adolf Meyer himself contended, a constructive school program was singularly important in establishing mental and moral health.[30]

Personal fulfillment, not social conformity, began to enter into concepts of positive mental health. Mental health was the result of constructive activities that were self-fulfilling and societally enhancing. Mental health was an active adaptation to one's environment, not a passive adjustment. Creativity, integration and independence were ideas frequently used. Those who championed mental hygiene still believed in the power of the individual to control his feelings and actions rationally. There was no escape from individual and social responsibilities. Mental hygiene and moral character were still intertwined inextricably.[31]

VI. CONCLUSIONS

For centuries, Western physicians have viewed the human body and its parts as real. Neither doctors nor patients have quibbled, in a metaphysical sense, about the locations of pain or disturbance in these parts. A stone is in a bladder and not an ear; the pain is at the site of a broken leg and not in a finger; the bullet is in the chest and not the foot; the swelling is in the abdomen and not the neck. Moreover, confusion, conflict, and nervousness are not extra-corporeal, although caused by emotional or interpersonal factors.

The bodily parts and mental capacities of an individual human undergo changes that produce discomfort, dis-ease. The nature of dis-ease can be understood by investigating the structural and functional alterations of the disturbed part or parts, or by investigating the abnormalities of mental and interpersonal processes. Using available techniques of observation and manipulation, countless scientists and physicians have been responsible for conducting these analyses, especially during the last five hundred years.

Neither have patients and doctors quibbled, in a moral sense, about the tasks of the doctor. He is responsible for alleviating the pain or removing the disturbance. In accomplishing these objectives, he is expected to do something, to perform therapeutic acts. Such acts will be rational if the nature and cause of the disturbance are understood (diagnosis). The previously reviewed studies of organismic and mental dis-eases outline the Western quest for this rationality. These investigations are still viewed as the primary obligations of medical scientists.

They impart rationality and justification to the therapeutic acts of physicians. A physician is responsible for assisting the healing power of nature, if such a power exists. A physician is responsible for altering the forces of nature, if nature is viewed as a never-ending series of structural and functional experiments. Without scientific knowledge of human bodies and minds, physician's judgments of diagnosis, prognosis, and therapy would be vacuous.

But, why should doctors be concerned about human suffering altogether? Because, diseases are not desirable; they are abhorred. With that value judgment, dis-eases become unworthy, de-humanizing. Congenital deformities, severe chest pain, cancer of the bowel – these abnormalities are not value-laden except in a negative sense. A well-formed newborn, no chest pain, a properly functioning bowel – these are valued. But they are healths, the opposites of diseases.

During the 18th, 19th, and 20th centuries, no classifications of healths have been created in the same way that classifications of diseases have been prepared. Physicians would probably belittle such a suggestion. Diseases are derangements of the structures and functions of the parts of an individual human body. If they are not deranged, they are healthy. Every textbook of human anatomy or physiology is believed to be a portrait of human health. Normality is the absence of abnormality; Ease is the absence of Dis-Ease. Without realizing it, perhaps, the mirror images of the elaborate nosologies of diseases are healths. The oppositeness is inherent, but seldom acknowledged as epistemologically different.

This basic difference between Dis-Ease and Ease resides in the fact that *ideals of health, not ideas of disease*, impart value to the meaning of therapeutics and hygiene. As previously indicated, the literature on hygiene – bodily or mental – is so impractical. On the one hand, it is a literature that has no meaning without reference to the prevention of diseases. On the other hand, it is a literature of ideals – exhortative, inspirational, humanistic. This idealism and its unitary tendencies are symbolized by the absence of a classification of healths, physical or mental.

One contemporary author claims that "mental health" is altogether an evaluative concept. "Mental health is personality evaluated, measured against certain criteria that either have the status of values or are derivatives of implicit values." [32] Such language is far-removed from language about kidneys, membrane potentials, viruses, and unconscious wishes. Through the centuries, students of organismic dis-eases and mental dis-eases have avoided value terminology in recording their observations and creating their explanations. Logically, though, they could never escape the value-ladenness of health, when they claimed that healths were the opposites of diseases. That there were as many healths as diseases was seldom acknowledged. Conceptual escape was possible because the idea of logical opposition was assumed, not explored.

In contrast, when these same clinicians reviewed their obligations as hygienists, either in studies of bodily hygiene or in studies of mental hygiene, they were unable to escape the value-ladenness of "health" concepts. They self-consciously included so many value judgments in their ideas of health and hygiene that, historically speaking, there were almost as many "healths" as there were human values.

In these ideals of healths, until recently, value primacy was given to the individual human. Yet, there is probably not in the entire world a complete medical record of the life history of a given individual human. Perhaps there is no more dramatic way to illustrate the basic problem in circumscribing human disease and human health. The practice of medicine has been a pragmatic matter of dealing with emergencies, of crisis care, focused on particular problems that are located in particular bodily parts or mental functions, and explained by dis-ease concepts. Few medical scientists have ever undertaken the systematic study of healthy humans who do not need their services, or the systematic study of patients as persons who experience alterations in their ways of living, alterations called dis-eases. The stark contrast between a biography and a hospital chart exemplifies the basic difference between personal problems and organismic dis-eases or mental dis-eases. This contrast also suggests that certain legacies in the philosophies of modern medical science need revision.

University of Texas Medical Branch, Galveston, Texas

NOTES

[1] For much of what follows in this section, I am indebted to Owsei Temkin. See especially

his essay on "Health and Disease," in *Dictionary of the History of Ideas*, ed. by Philip Wiener (New York: Scribners, 1973), II, 395–407.

[2] For more details about the 17th and 18th centuries, see the following books by Lester King: *The Road to Medical Enlightenment, 1650–1695* (New York: American Elsevier, 1970) and *The Medical World of the 18th Century* (Chicago: University of Chicago Press, 1958).

[3] Guenter B. Risse, "The Quest for Certainty in Medicine: John Brown's System of Medicine in France," *Bulletin of the History of Medicine* **45** (1971), 1–12. Another ontological fallacy could be called synecdochic. This one was not infrequently committed by arguing that disturbance in a certain part of the body was identical with the dis-ease of the whole individual human.

[4] Knud Faber, *Nosography in Modern Internal Medicine* (New York: Paul B. Hoeber, 1923), pp. 112–171.

[5] Esmond R. Long, *A History of Pathology* (New York: Dover, 1965), pp. 89–168.

[6] "Nomenclature of Disease," *Transactions of the American Medical Association*, 1872, Appendix in Vol. 23; *The Bellevue Hospital Nomenclature of Diseases and Conditions* (New York: Board of Trustees of Bellevue and Allied Hospitals, 1911); United States Public Health Service, *Nomenclature of Diseases and Conditions* (Washington: Government Printing Office, 1921); *Standard Nomenclature of Diseases* (Chicago: American Medical Association, First Edition, 1933); and subsequent editions of the *Nomenclature* published by the American Medical Association.

[7] Robert Montraville Green, *A Translation of Galen's Hygiene* (Springfield: Charles C. Thomas, 1951).

[8] Owsei Temkin, "What is Health? Looking Back and Ahead," in *The Epidemiology of Health*, ed. by Iago Galdston (New York: Health Education Council, 1953), pp. 15–18.

[9] Hebbel E. Hoff and John F. Fulton, "The Centenary of the First American Physiological Society Founded at Boston by William A. Alcott and Sylvester Graham," *Bulletin of the History of Medicine* **5** (1937), 688.

[10] *Transactions of the American Medical Association* **1** (1848), 272–274.

[11] William B. Carpenter, *Principles of Human Physiology, With Their Applications to Pathology, Hygiene, and Forensic Medicine*. First American edition by Meredith Clymer (Philadelphia: Lea & Blanchard, 1843), p. 27.

[12] John Bell, *Health and Beauty* (Philadelphia: E. L. Carey & A. Hart, 1838), p. 18.

[13] *Ibid.*

[14] A. H. Buck (ed.), *Hygiene and Public Health* (New York, 1879).

[15] *The University of Texas Medical Branch at Galveston, A Seventy-Five Year History by the Faculty and Staff* (Austin: University of Texas Press, 1967), pp. 78, 88, 110; also, see Victor C. Vaughan, "Methods of Teaching Hygiene," *Philadelphia Medical Journal* (September 1, 1900), 401–404.

[16] E. Stanley Ryerson, "Health and Medical Education," *Journal of the Association of American Medical Colleges* **13** (1938), 1–13.

[17] Donald G. Bates, "Medicine's Role in Comprehensive Care," *Perspectives in Biology and Medicine* **15** (1972), 317–325; John G. Bruhn, "The Diagnosis of Normality," *Texas Reports on Biology and Medicine* **32** (Spring, 1974), 241–48; Alfred H. Katz and Jean Spencer Felton (eds.), *Health and the Community Readings in the Philosophy and Sciences of Public Health* (New York: The Free Press, 1965).

[18] Paul W. Harkins, *Galen on the Passions and Errors of the Soul* (Columbus: Ohio State University Press, 1964).

[19] As an appendix to *The Vital Balance* (New York: Vikings Press, 1963), Karl Menninger

prepared an extraordinary review of nosologies in the history of psychiatry. For Paul of Aegina, see p. 423 and for Aquinas, see p. 425.

[20] *Ibid.*, p. 432.

[21] Thomas C. Upham, *Outlines of Imperfect and Disordered Mental Action* (New York: Harper and Bros., 1840).

[22] Menninger, *The Vital Balance*, pp. 454–457.

[23] Henry J. Berkley, *A Treatise on Mental Diseases* (New York: D. Appleton, 1900); Smith Ely Jelliffe and William A. White, *Diseases of the Nervous System* (Philadelphia: Lea & Febiger, 1915).

[24] See note 6.

[25] This is argued explicitly in the numerous writings of Thomas Szasz.

[26] Isaac Ray, *Mental Hygiene* (1863; reprint ed., New York: Hafner Publishing Co., 1968), p. 15.

[27] Barbara Sicherman, *The Quest for Mental Health in America, 1880–1917* (Ann Arbor: University Microfilms, Inc., 1971), pp. 78–152.

[28] Paul V. Lemkau, "Notes on the Development of Mental Hygiene in The Johns Hopkins School of Hygiene and Public Health," *Bulletin of the History of Medicine* **35** (1961), 169–174.

[29] William A. White, *The Principles of Mental Hygiene* (1917; reprint ed., New York: Arno Press & The New York Times, 1972), p. 34.

[30] Adolf Meyer, "Mental and Moral Health in a Constructive School Program," in *The Collected Papers of Adolf Meyer*, ed. by Eunice A. Winters (Baltimore: The Johns Hopkins Press, 1952), IV, 350–370.

[31] Sicherman, pp. 329–410.

[32] M. Brewster Smith, "'Mental Health' Reconsidered: A Special Case in the Problem of Values in Psychology," *American Psychologist* **16** (1961), 673; for a more recent assessment of "mental health," see Ronald Leifer, *In the Name of Mental Health: The Social Functions of Psychiatry* (New York: Science House, 1969).

SECTION II

PHILOSOPHY OF SCIENCE
IN TRANSITION TO A PHILOSOPHY
OF MEDICINE

STEPHEN TOULMIN

CONCEPTS OF FUNCTION AND MECHANISM
IN MEDICINE AND MEDICAL SCIENCE

(Hommage à Claude Bernard)

If we are to make the transition from the philosophy of science to a full-fledged philosophy of medicine, our first task will be to find a way of distinguishing between medicine and science; and of doing so in completely general terms. How is this to be done? And, once it is done, how are we then to describe the relations between the two enterprises, again in completely general terms? Is a knowledge of medicine something different in kind from a knowledge of the sciences whose results find application in medical practice? Or does the difference between the two enterprises, such as it is, lie in the nature of the applications to which that knowledge is put, rather than in the knowledge itself? When, for instance, Claude Bernard published his great methodological treatise 109 years ago, and called it *An Introduction to Experimental Medicine*, was this title appropriate or was it a misnomer? Should he have called it instead, say, *An Introduction to Experimental Physiology as Applicable to Medicine*?

Questions such as these are fundamental to any fruitful discussion of today's topic. In actual practice, the clinical enterprise of medicine is clearly interlinked with the scientific enterprises of anatomy, biochemistry, physiology and the rest, in an intimate, complex, even inextricable manner. By now, we see at most shifts of emphasis within the whole activity of medicine, between the clinical and scientific aspects of medical work, but we no longer hear any suggestion that medicine and science could ever divorce. After a decade of public preoccupation with the so-called "biomedical sciences," for instance, we have recently seen the focus of medical attention shift to the topic of "health care," with a corresponding tendency to retreat from the scientific side of medicine. But the social and political importance of delivering the fruits of medical knowledge to all who are in need – rather than just to those for whom the Mayo Clinic (or Mass. General, or whatever) is within geographical and financial reach – does not affect the deeper questions, in what this medical knowledge consists, and how it stands with respect to the relevant branches of natural science. These are the questions I shall be concentrating on here.

H. T. Engelhardt, Jr. and S. F. Spicker (eds.), Evaluation and Explanation in the Biomedical Sciences, 51–66.
All Rights Reserved. Copyright © 1975 by D. Reidel Publishing Company, Dordrecht-Holland.

There is one familiar and respectable line of attack on these questions that has some real attractions, and puts us apparently in a position to draw a sharp and clear distinction between the two kinds of enterprise. This is to say that Science is essentially *value-free* and that Medicine is essentially *value-laden*. Medicine being practical, and directed towards human good, a knowledge of medicine is (on this view) a grasp of practical maxims, all of which are in effect *hypothetical* imperatives: that is, directions about how we are to proceed *if we wish* to achieve the goals, or realize the goods, that are consecrated in the physician's fundamental ethical commitments. By contrast, the sciences on which medical practice relies are (so the argument continues) as ethically neutral, in themselves, as any other sciences. They are indifferent as between human preferences, even as between human good and ill. They simply study the correlations between physiological phenomena, the rates and products of biochemical reactions, law-governed processes, mathematical relationships and so on; and, while they may contribute greatly to our understanding of the bodily happenings that are the physician's immediate concern, they help us to understand both the good and the ill of medicine – both the good of health and the ill of disease – equally and in the same terms. In brief: medicine is essentially concerned with "healthy functioning," as contrasted with "diseased malfunctioning," and both of these are evaluative notions that take a particular choice of values for granted. Physiological science, on the other hand, is concerned with biophysical and biochemical "mechanisms," which are value-neutral, being called into play indifferently in health and disease, for good or for ill, within the healthily functioning organism and within the diseased or malfunctioning one.

For all its attractions (I shall argue) this position is mistaken. This is not to say that it is downright *false*. It makes a point, and the point may be worth making in some contexts. But it makes this point in an oversimplified and misleading way, and in doing so it conceals more about the actual relations between science and medicine than it reveals. In particular, it relies too absolutely on a distinction between "functions" and "mechanisms" that is more easily stated in words than applied to actual cases.

If we set this over-sharp distinction aside, however, and take a closer look at the relations between the functional and mechanistic aspects of the organism, we shall be led to another, more complex and richer view.

According to this alternative view, there is no clear division of natural processes in the real world, into "functions" on the one hand and "mechanisms" on the other. Rather, we draw a distinction between the functional and mechanistic *aspects* of any natural process, in one context or another, from one point of view or another; and whatever can be viewed as a mechanism, from one point of view and in one context, can alternatively be seen as a function, from another point of view or in another context. Indeed, the very *organization* of organisms – the organization that is sometimes described as though it simply involved a "hierarchy" of progressively larger structures – can be better viewed as involving a "ladder" of progressively more complex systems. All of these systems, whatever their levels of complexity, need to be analysed and understood in terms both of the functions they serve and also of the mechanisms they call into play. And, when we shift the focus of our attention from one level of analysis to another – from one fineness of grain to another – even those very processes which began by presenting themselves to us under the guise of "mechanisms" will be transformed into "functions."

This alternative view (as we shall see) undercuts any attempt to enforce a sharp distinction between value-neutral science and value-laden medicine; and it confirms the justice of the insights that Claude Bernard enshrined in his theories. In his great book, he was indeed writing about the vital processes of the organism – part-mechanisms, part-functions – which are the joint concern of the physiologist and the physician, and so about *Experimental Medicine*. And, behind his analysis, we can already discern, in outline, the philosophy of medicine that (as I see it) underlies the contemporary union of science and medical practice; a philosophical attitude that one might call *Neo-Stoicism*. In this as in so much else, *hommage à Claude Bernard*.

Let me begin my argument by recalling Bernard's key contribution to physiology. This was his recognition of the principles of operation of homeostatic mechanisms in organisms: specifically, the role of the vasomotor system and similar systems in regulating the crucial properties of the circulating fluids in living organisms – those fluids that play so large a part in determining what he called the *milieu intérieur*, or "internal environment." Reading Claude Bernard with late-twentieth-century eyes, at the price of slight anachronism, we can already see in his argument a vision of the organism's unity and functioning as maintained, so to speak, by the "convection of information" through this internal environment; with a hundred mechanisms (electrophysiological, biochemical, immunological and so on) ready and set to respond correctively to discordant or deviant signals. Although when Bernard was writing, a century and more ago, such men as Cannon and Henderson, Shannon and McCulloch, were all in the future, their analysis of homeostasis and feedback is already prefigured with some exactness in Bernard's own work.

One thing especially Bernard understood with notable clarity: the relationship between the *universal* regularities and laws of nature that are the concern of physics and chemistry, and the *particular* regularities and laws of bodily function that are the business of physiology. Before he set our minds at rest, the co-existence of physiology with physics and chemistry had always given rise to perplexity. Equating the laws of physiology with those of physics and chemistry seemed to involve denying the special features that mark organisms off from non-living things; yet to insist on those special features at any cost seemed to involve exempting organisms from the universal sway of physico-chemical law. Neither position was satisfactory to any except ideological enthusiasts, and the very assumption that these are the only possibilities open to us was, as Bernard saw, a misconception. The regularities we study in physiology spring in themselves, he argued, neither from the conformity of organic processes to the laws of physics and chemistry – since that would fail to distinguish organic phenomena from inorganic – nor from any power in organisms to override those laws, since that would make physiological phenomena physically and chemically unintelligible. They originate in the special character of the local *micro-environments* existing in the human and animal frame, within which the working-out of general physical and chemical principles is capable of giving rise to specifically *physiological* regularities. This recognition, that the special conditions existing locally within living organisms are sufficient to explain the characteristics of "vital processes," without the necessity either to deny their unique features or to exempt them from the operation of physics and chemistry, was what led Bernard to call himself a "physical vitalist." Physiology was not in opposition to physics and chemistry. It was simply physics and chemistry carried out "in the special field of life."

Taking Bernard's analysis as our starting-point, we can develop a more general account of the relations between function, mechanisms and causality in medicine and medical science. For this purpose, we need to make a series of distinctions aimed at clarifying the differences between the "functional" and the "mechanistic" aspects of organisms and physiological processes. The first two of these distinctions are preliminary, ground-clearing ones: lemmas, so to speak, which need to be dealt with before we can make explicit the third, and most significant of them.

To begin, then, with the most obvious and familiar point: whenever we distinguish between a bodily function and the mechanisms it calls into play, we are contrasting some physiological *outcome* – the body's temperature remaining constant, newly-eaten food being digested – with the biophysical and biochemical *workings* whose occurrence brings about this outcome. Mechanisms serve functions; and they must be capable of bringing about the outcomes that constitute the proper maintenance of the functions in question. No doubt, the same mechanisms are also capable in other circumstances, e.g., if the stability of the internal environment is upset, of bringing about those alternative outcomes that constitute *mal*function or *dys*function of the organ or system concerned. But the criteria for deciding what represents a proper function, and what a malfunction, are normally specifiable at the outset in terms prior to, and independent of, any detailed knowledge of the mechanisms called into play in the process. (True, we often sharpen up those criteria subsequently, in the light of our understanding of the mechanisms involved; but that is another matter. The initial description of the function can commonly be given, in gross terms, without reference to those mechanisms: e.g., by remarking that perspiration evaporates, dissipating heat, so that the body maintains a constant temperature, or whatever.)

The second preliminary point has to do with two distinct applications of the term "function." On the one hand, we use the term to refer to the operation of some localized organ: specifically, to the occurrence of some local outcome. (The liver secretes bile, the pores dilate and contract, and so on.) Clearly, the most exact specification of these *local functions* comes close to being an account of the bodily mechanisms involved; and improvements in our knowledge of those mechanisms readily lead to changes in our characterizations of the local functions themselves. In this sense, the question, "What is the function of the liver?," is not easily separated from "What does the liver do?," or from "What happens in the liver?" On the other hand, the term "function" is also used in a broader sense, to refer to the role of some particular organ, or system, in the entire life of the organism, and so to the overall outcome of its proper functioning or malfunctioning. (Digestion, respiration and reproduction are "functions" in this sense.) Evidently, a great deal was known about these overall *vital functions* before anyone had any clear picture of the mechanisms called into play in them. All sorts of smaller organs, organelles and mechanisms, for instance, serve to maintain these vital functions, and the local functions of the smaller parts or constituents contribute to the successful maintenance of the corresponding vital functions; but what these smaller parts all are, and how they contribute, is much more recent knowledge than the basic understanding of the vital functions themselves.

If we are to reconsider and refine our understanding of vital functions, indeed, this will come about, not so much as a result of any better knowledge of the organism's inner processes and mechanisms, but rather from a more subtle analysis of its whole life, and of its evolutionary ancestry. On this level real discoveries are still being made about (e.g.) the perceptual function of eye-movements and the like. Yet, what, for instance, is the function of blinking? I'm not sure that we really know.

One comment in passing: though this second distinction, between local functions and vital functions, is easily acknowledged when it is made explicit, we often talk in actual practice about the "functions" or organs and systems in a way that glosses over that difference. In our normal descriptions, for instance, local functions tend to be superseded by the larger-scale vital functions to which they contribute. After all, the local outcome of a particular organ's *modus operandi* has to be as it is, if it is to be successful in serving that larger-scale function; and many smaller interdependent parts combine together into the more complex systems that directly serve the vital function in question. They act in the ways they do, as Aristotle would have put it, "for the sake of" the larger vital function. So, while the kidney or pancreas operates in a way that maintains its own local systemic integrity, we would not be inclined to say that either of them had "a good of its own," unrelated to the vital function of the entire digestive or eliminative system. If anything, the tendency of a carcinoma to pursue "its own good" independently of the rest of the body is the feature that makes it pathological.

(Incidentally, this point has some relevance to the use of physiological analogies in talking about society. Though Durkheim may, for instance, be right to emphasize the interdependence of individuals in a social group, the way in which he does so risks suggesting that the "good of his own" that each individual citizen might otherwise pursue should be subordinated to a problematical "good of society" that is to be pursued collectively, without being to anyone-in-particular's good. Whereas, presumably, each individual citizen really *does* have "a good of his own," in a way that the individual bodily limb or organ does not; and our interdependence in society implies, as a consequence, that my pursuit of my personal good should be qualified by respect for your pursuit of yours, rather than superseded by some imagined organic counterpart, in the form of an overall "collective social function.")

So much for preliminaries: now let me turn to my third distinction, which brings us to the heart of the matter. Suppose that we take Claude Bernard's account of a local function, as involving a physico-chemical process which gives rise to a physiological regularity, through taking place in a special organic micro-environment; and suppose that we try to match the different elements in this complex account to our naive categories of "cause" and "effect"; in that case, we shall at once find ourselves in a quandary. For there are several different ways of doing this matching, there is no obvious way of choosing between them, and there is a good deal of potential confusion in the offing as a result. On one level, we may be inclined to apply the term "cause" to the *physico-chemical* ingredients and factors (e.g., food substances) entering into the functional process in question; on another, to the special features of the micro-environment (the large intestine, say) that serves as the *locus* of that process; on a third, to the occurrence which is the *occasion* for the function (eating stimulates the digestion). Correspondingly, we may be inclined to reserve the term "effect" for the functional *outcome* of the digestion, say, or alternatively for the physico-chemical *products* of the associated mechanisms.

With a minimum of three categorially different kinds of cause and two kinds of effect, there is evidently no single, simple way of characterizing the *causality* of such a functional process, and much ink can easily be spilt arguing about the matter. For instance, does ordinary physical causality apply to organic processes? The answer depends on what you mean by the word "apply." As Bernard insisted, the things that happen on a biophysical or biochemical level *in vivo* should be assumed to be, in this respect, no different from the corresponding physical and chemical processes happening *in vitro*. Both conform to the same general laws. But the physiological regularities that link occasions and functional outcomes, in specific organic loci, illustrate a different kind of causality: one much more like the "causality" of everyday life as discussed, say, in the lawcourts. The question at issue in that case is not, "What universal law does this process exemplify?", so much as, "What conditions are required to ensure, or frustrate, the successful outcome of this functional process?"

To avoid confusing the causality of physiological functioning with that of physico-chemical processes, it will be best here to skirt around the over-simple words, "cause" and "effect," and to use a more complex classification of factors in their place. Thus, on the level of mechanisms, we may speak of "ingredients" and "products," "pressures" and "forces," "actions" and "reactions;" on the level of functions, we may speak of "loci," "occasions" and "outcomes" (or "responses"); and we must be prepared to find people linking several different pairs from these groups as "causes" and "effects." In addition to purely physico-chemical linkages of action and reaction, and purely functional linkages of occasion and outcome, for instance, we shall also find – to compound confusion – cross-linkages between mechanistic factors and functional outcomes. (When a man inhales car-exhaust fumes, the bonding of carbon monoxide molecules to the haemoglobin in his blood may thus be said to "cause" the resultant asphyxiation. In this case, the connections between the legal and medical uses of the term "cause" are quite patent.)

Bearing in mind this last distinction, between the half-dozen-odd kinds of factors involved in any functional process, we are now ready to return and deal with the central thesis of this argument. That thesis had to do (you recall) with the relationship between the functional and mechanistic *aspects* of physiological phenomena. Organic processes in the real world (I claimed) do not fall naturally, of themselves, into separate classes of functions on the one hand, mechanisms on the other. Whether we shall view any particular process as a "function," or alternatively as a "mechanism," will depend rather on the context of inquiry; and one and the same process can be viewed in either of these lights, depending on the context and the point of view. Why this is so, can be appreciated much more clearly, if we restate the point using the distinctions I have remarked on here.

To begin with, then: (1) In a typical physiological process, the factors that we would normally speak of in terms of physical causality and general laws – as "actions" or "reactions," "ingredients" or "products," and the like – all of them lie *within* the associated physico-chemical mechanisms. By contrast, the "occasion," the "locus" and the "outcome" of the process are all of them functional factors, *external to* the mechanisms that are called into play to serve the function in question. At the same time, however: (2) The immediate local functions served by these physico-chemical mechanisms commonly manifest small-scale physiological regularities, in which "occasions" and "outcomes" are correlated in a simply intelligible manner. Where this is the case, the natural shorthand of scientific discourse may lead us to construe those very correlations as straightforward sequences of "cause" and "effect," and even to speak of them in turn as "mechanisms."

When they are considered as contributing to the operation of the liver and bile-duct, for example, the biochemical transformations going on in the liver are the "mechanisms" called into play in the "function" of bile-secretion; but bile-secretion itself – particularly, the variation of bile-secretion with changing stimuli and occasions – can itself be viewed as a "mechanism," when it is considered as contributing to the "functional" operation of the entire intestinal tract; and this latter operation in turn becomes a "mechanism," as seen from the standpoint of the overall "vital function" of digestion. ... The localized functions of individual organs and systems are thus interrelated, so that the larger-scale systems perform more complex functions, and the systemic integrity and regular operation of the constituent smaller-scale systems will then be a precondition for "healthy" outcomes from the larger-scale systems they make up. So, we may treat the vitamin-C metabolism, say, or the operation of the vaso-motor system, or the stabilization of the visual image, *either* as the functional "outcome" of smaller-scale mechanisms, *or else* as a "causal" mechanism that is itself called into play to "effect" some more complex outcome. In the first case, we speak of it as a "function" associated with a smaller-scale "mechanism;" in the second case, as the "mechanism" serving a larger "function."

This picture of larger and smaller-scale *systems*, in which more or less complex mechanisms serve more or less complex functions, is the ultimate contemporary destination that we have arrived at in physiological theory, as a result of following up the developments originally initiated by Claude Bernard. From this picture, it is clear that the comparative spatial dimensions of different organs and bodily structures is connected only incidentally with their comparative complexity as "functional systems." The different *levels of organization* of different physiological systems are a reflection of the respective *degrees of complexity* of the functions they serve, rather than of their mere size and scale.

It is true, of course, that the more complex systems will tend to be larger in scale than the less complex systems they incorporate, if only because they *do* incorporate them, physically as well as functionally. Yet, certainly, the most massive and bulky organs in the body are not necessarily the most complex. It is also true that the more complex systems will tend to operate on longer time-scales than the less complex ones that are their constituent mechanisms; if only, again, because they *are* dependent on those other smaller and quicker systems to maintain their regular functioning. But, here too, the slowest-acting processes in the organism are not necessarily those that serve the most complex functions. By most intuitive standards, indeed, certain neurophysiological systems appear among the most subtle and complex, both in their multitude of pathways and in their modes of functioning; yet those neural systems are compact and quick-acting in comparison with many other organs and systems, such as those involved in the digestion. So, in the end, we are left with a conception of the "organism" and its "organization," not so much as made like a kind of Chinese box, with larger and smaller structures packed one within another, but rather as dependent for its "vital functions" on a ladder of more or less complex systems, with more or less complex "local functions;" and, as we pass up or down this ladder from one step to another, what are labelled as "functions" on the lower step become "mechanisms" from the vantage-point of the next higher level.

The existence of such physiological systems, on different levels of complexity, has been a factual discovery for clinical medicine and medical science alike; and it is one that is taken for granted in most subsequent interpretations. Neither physicians nor scientific physiologists now question the existence of the individual steps in the ladder of functions. Instead, they are concerned to explore the patterns of systemic relations between the processes going on at different levels of this ladder. Having arrived at this point, they take it for granted that the integrity of the organs and systems involved on each level is "a good thing;" and this assumption too is made equally in both clinical medicine and medical science. Neither in physiology nor in medical practice, for instance, do we have any occasion to ask whether the continued existence and functioning of the heart and circulatory system are "desirable." In this respect, John Stuart Mill was clearly right. So far as this sort of health is concerned, the "desirable" is one and the same with what is actually desired; and no more fundamental maxim can be appealed to, in order to underwrite any further the belief that a healthy heart and circulation are "good."

There is, correspondingly, very little substance to the idea that a scientific physiology could, even in principle, be "value neutral." The chief vital functions of the human body are not merely "good in themselves." They are preconditions for almost any other imaginable human good. As for the lesser systems and the more local functions that cooperate to maintain those vital functions: their desirability is evident simply from their status as the instruments of good health. So it is no longer the case that the physician has nowadays to make any specifically ethical commitment to sustain health, or to do whatever happens to be, physiologically speaking, conducive rather than opposed to that end. His whole enterprise is nowadays intelligible, and capable of being rationally expounded, only in terms of a systemic picture of vital organization and functioning; and this picture presupposes the existence and integrity of all the main functioning systems in the human frame.

The same systemic picture forms the core of modern physiology also. If the basic empirical subject-matter of that science were composed merely of biophysical and biochemical processes, directly explicable in terms of general laws and physical necessities, we might perhaps declare physiology indifferent as between the evaluative conceptions of "health" and "disease." Instead, modern physiology is intelligible only in terms of the same "organic systems" that lie at the base of medicine: systems whose existence and character are defined in terms of the functions they serve.

As a result, the logical gulf between "facts" and "values" has little practical significance for either medicine or medical science. The fundamental values of clinical medicine are rooted directly in the basic facts about vital organization, while the basic facts of physiology can themselves be stated only in terms that take for granted the value of the chief vital functions of the organism. Either way, the *rational* significance of vital organization or functioning is inseparable from its *ethical* significance.

It is this intimate union of facts and values, in a conception of "organic systems" that dovetails "function" and "mechanism" completely, that leads me to compare the philosophical presuppositions of contemporary physiology and medicine with those of the Stoics. For 19th-century mechanists, the task of fathoming the operations of the human body was no different in method or spirit from that of understanding the make-up and working of any other physical object. Its complexity was merely *quantitative.* If the healthy working of the body was also perceived as "a good thing," that was an expression of our choices, our preferences, our values. For us today, the mere number of individual atoms and cells in the body is a secondary matter. Its complexity is chiefly systemic and functional; and, to the extent that its value and intelligibility alike spring from that same systemic character, we have no choice whether to prefer normal to pathological working, function to dysfunction, health to disease.

In some respects, of course – in neurophysiology, for instance – the *degree* of this systemic character has no inorganic counterpart. Until Fred Hoyle's "black cloud" turns up in real life, we may expect to go on finding full-scale self-awareness and rationality associated only with organic systems of a neurophysiological order. Still, this difference of degree does nothing to cut off the organization of the human body from the rest of Nature. On the contrary, we might say: at every level of complexity, *Nature comes in Systems.* There is sufficient continuity between the organization of the atom and those of the macromolecule, the cell and the organism, for us to apply many of the same concepts and categories right across the board.

Indeed, the development of sub-atomic physics during the twentieth century has led to the same kind of changes in our ideas about physical systems that the change from 19th to 20th-century physiology has produced in our ideas about living organisms. By contrast with the random world of, say, Maxwell's kinetic theory of gases, the order and harmony that reign in the quantum world of stable orbits and eigenstates has transformed the atom into – forgive the phrase – a kind of "inorganic organism." The sodium atom, the water molecule, the benzene ring: on an elementary enough level, each has its own systemic integrity and *modus operandi*, as much as the cell, the kidney or the Krebs cycle. Spatial arrangement is no more the key feature of an eigenstate than it is of an organic system. Rather, the alternative eigenstates of an atomic system represent alternative modes of functioning or "harmonies." (The Stoic term is a natural one in this context.) So the quantum world, too, differs from the classical Newtonian world in being a world, not of structures, but of systems.

From the point of view I have been arguing here, the intellectual unity of medicine and physiology is effectively complete. The difference between the values of the two enterprises does not lie in any intrinsic commitment to "facts" (or "mechanisms") in the one case, "values" (or "health") in the other. In both enterprises we are dealing with a similar compound of mechanisms and functions, facts and values. The difference lies, rather, in their respective attitudes to the external application of that common understanding. We have reached a point at which somatic medicine has become, intellectually speaking, very largely applied physiology and biochemistry.

This conclusion leaves two further questions unanswered, and I shall say something very briefly about them in conclusion. First: (1) I have restricted my conclusion to *somatic* medicine, and the question naturally arises, how far my analysis lends itself to including mental as well as bodily health and disease. This is not a simple question to deal with. The gap between Claude Bernard's rather static account of physiological function, and the notions we need in order to understand psychological development and functioning, is a broad one. For a start, Bernard concerned himself primarily with homeostasis, and with the role of the *milieu intérieur* in maintaining the equilibria essential for normal bodily functioning; and he was himself openly skeptical about the chances of extending this treatment to such developmental topics as embryology or morphogenesis, to say nothing of psychology. By now, we may be more confident than he was that normal morphogenesis, like normal physiological functioning, is controlled by intelligible mechanisms; so that the equilibrium concept of "homeostasis" can have its developmental counterpart – what we might call "orthokinesis." And it is tempting to go on and suppose that psychological development may in due course yield to a similar analysis, involving perhaps an external *milieu culturel* or *milieu éducatif* as much as the physiological *milieu intérieur*.

This program has, however, certain important limitations. There may be many respects in which cognitive development represents, so to say, a "realization of innate capacities" – more respects, perhaps, than we have yet had the opportunity to discover. To that extent, the problem of "psychic health" will indeed be one of finding out the conditions on which a child can achieve its full mental, as well as its full physical stature. Yet we must all be familiar with the dangers of turning the phrase "mental health" into a metaphorical catch-all. One need not refer to the totalitarian use of mental wards as an instrument for suppressing political dissidents: there are similar risks to be faced much nearer home. For instance, we all know how hard it is to draw sharp lines at the boundaries dividing outright mental disease from emotional disturbance, lack of judgement, or even from sheer misunderstanding of one's personal situation; and, correspondingly, how difficult it is to distinguish psychiatric diagnosis at the boundaries from psychotherapy, counselling, or sheer friendly advice and reassurance.

The dangers built into the metaphorical uses of the phrase "psychic health" are, in fact, the same as those built into the "organic" theory of society and the State. They can be used to provide an excuse for imposing conformity in areas of conduct where we are in fact entitled to a genuine choice in our courses of action and styles of life. So understood, indeed, the metaphors of "psychic health" and of "the social organism" support and reinforce one another. Both of them treat social relations as analogous to organic connections, and depict the continued integrity of a particular set of social forms as self-justifying, like the continued integrity of, say, the central nervous system. So a man whose style of life disregards the supposed "collective good" of society is not tolerated, or even respected, as being an independent, autonomous individual, but is liable to be denounced as being "sick" or "disturbed." In extending our discussion from physical to mental health we must be on the watch, here as ever, against the misuse of medical categories as instruments of social and political oppression.

(2) My other concluding question is also, by implication, a social, and even a political one. If we are to make the transition from the philosophy of science to a fully-fledged philosophy of medicine, our first task (as I said at the beginning of my remarks) is to find an adequate way of distinguishing between science and medicine, and of doing so in sufficiently general terms. My subsequent argument has done little, I fear, to help with that problem: at most, I have closed off certain lines of attack on it, by insisting (e.g.) that the difference between clinical medicine and medical science is *not* adequately described by labelling the first "value-laden," the second "value-free." So far as the understanding embodied in the two enterprises is concerned, both are equally committed to the same "systemic" analysis, and an acceptance of Nature as System brings with it for us today – as it did for the Stoic philosophers – an implicit body of values. The nature of *health* is, at one and the same time, a matter for empirical discovery and a matter of evaluative decision. We refine our sense of how the human body *ought to* work, and ought to be *helped to* work, in the course of and in the light of our empirical studies of how it *does in fact* work. The question, how we should care about and nurture the body, may thus be distinguishable from the question, what things in fact contribute to the workings of the body, but the two questions are nonetheless inseparable. In medicine, as in law, we face choices and decisions that can be made conscientiously, only in the light of the best empirical understanding we can command; and, once it is sufficiently complete, this empirical understanding commonly leaves us precious little room for real choice.

If the distinction between medical practice and medical science is not to be characterized by pointing to the kinds of knowledge typical of each enterprise, in what direction are we then to look for that difference? The answer to that preliminary question I have already hinted at, when I spoke of medicine as being largely "applied physiology;" a point I might have done better to put in different words, by reversing the order of terms, and saying that *physiology* is the purely theoretical, or "scientific," part of *medicine*. So, let me ask: if we start with the living totality of medicine, and make a move to the abstraction called "physiology," what do we leave out in the course of that abstraction? The answer is: all concern with values and choices *other than* those intrinsic to the concepts of medical science itself – all concern with questions about the social context of medicine; about priorities in the availability of health care; about whose welfare our accumulated medical understanding is to be applied to promote, on what conditions, for what price; about the political role of the medical profession, and so on....

That is a group of topics we shall have the chance of discussing this evening; and it is one that takes us far beyond the first, prosaic considerations about the concepts of "function" and "mechanism" that I have been discussing here. For my own part: if I have done something to encourage doubts in your minds about the commonplace picture of natural science – including medical science – as a purely factual, value-free enterprise, and to remind you of some of the ways in which our empirical understanding of the world, and the concepts in which that understanding is embodied, brings implicit values with it, that will have been enough for one paper.

University of Chicago,
Chicago, Illinois

MARX W. WARTOFSKY

ORGANS, ORGANISMS AND DISEASE:
HUMAN ONTOLOGY AND MEDICAL PRACTICE

At the end of his fine essay, "The Diseases of Civilization," René Dubos
writes

The disorders of the body and the mind are to a very large extent the consequences of
inadequate responses to the environment. They involve not only a particular organ but
the organism as a whole. For this reason, the practice of medicine demands of the physician a holistic attitude that goes beyond that of the experimental scientist.[1]

My argument in this paper is that one has to go beyond the whole organism as well, and that the locus of medical practice extends to the
social, historical and cultural contexts of good health and ill-health. It
is my view that the ontological domain of medical practice is the socio-
historical ontology of the human being – not of the biological "whole
organism" as such, but of what constitutes this organism as a human
organism; that is to say, what goes beyond the merely biological characterization of organisms to the species-specific characteristics of human
beings. In general, I would argue, the analytical abstraction of the human
as a biological organism of a specific type leaves out what is most essential in differentiating the human as subject (both of medical practice
and of other practices as well – educational, political, cognitive), namely,
the crucial constitutive features of human sociality, human historicity,
and the concomitant self-identification *as* human which comes with linguistic communication and culture.

However, to make my point more specific, with regard to human
ontology and medical practice, I am going to focus here on what I take
to be the ontology of medical practice – namely, that conception of the
domain of medical practice which reflects the practitioner's view of the
reality he deals with, or of the subject of his practice. Here, I will argue
that it is not the *organ*, nor yet the *organism* which is the appropriate
basic entity in that domain, but rather the *disease*; and that the appropriate characterization of the disease is precisely the one which takes it
as a socio-historical and cultural phenomenon. In effect, my strategy is
to argue from what I take to be the appropriate characterization of human

H. T. Engelhardt, Jr. and S. F. Spicker (eds.), Evaluation and Explanation in the Biomedical Sciences, 67–83.

ontology to the isomorphic and appropriate characterization of the ontology of medical practice.

The critical thrust of the paper is therefore to attack both the theoretical and practical attitudes which reduce the domain of medical practice either to the organ or the organism alone. The constructive aim of the paper is to propose a modified holism – what I would call an emergent holism – in which the specific characters of the organ and of the organism are reinterpreted in the light of the socio-historical and cultural contexts of human ontology, and of a parallel definition of disease.

The aim of the paper is therefore normative – both critical and constructive. But one could proceed in a different way. That is, one could start with medical practice as it is, descriptively, and reconstruct what the alternative existing ontologies are which are entailed or suggested by these actual practices. The question addressed in this way would be: Do alternative modes of medical practice and medical theory constitute the objects of treatment in different ways? To put it differently: *is the subject of medical practice an artifact of the mode of treatment*? The point of the question is to place the very practice of medicine in a critical context, and to consider, in a philosophically and methodologically reflective way, how different traditions, styles, and canons of medical theory and practice pick out different entities as crucial for consideration. Such alternative sets or types of entities thus constitute different ontologies – different constructions of reality. In this sense, one could say that medical ontology recapitulates medical methodology, and a descriptive account of what physicians take their world to be, as practitioners, may very well reveal not one, but many alternative, and perhaps incommensurable ontologies. If one considers the complex institution of medicine as a whole, including medical education, and the varying levels of medical practice both in personal and institutional contexts – e.g., the private practice, the clinic, the hospital, the community health center, the research laboratory – and in the various specialties, and within different frameworks and levels of approach – e.g., epidemiology, infectious diseases, ethnic or genetic diseases, historically extinct diseases, somatic diseases of individual organs, etc. – then the alternative "entities" which constitute the *basic* domain in each of these cases will be seen to differ sharply.

Such a descriptive account of the "dispersions" which are revealed in actual practice would indeed constitute a kind of Foucaultian history of medicine (if not an "archaeology" of medicine). But the strategy of my paper is not critical in *this* sense. I am not concerned with the historicity and the relativity of actual medical practice in all of its details. No doubt such an account would be useful, especially in making medicine historically and epistemologically self-conscious. The injunction here would be, not "Physician, heal thyself," but, in a different therapeutic context, "Physician, know thyself." Holding the mirror up to medical practice is, however, not yet critical in a normative sense. Therefore, my own critical strategy is to propose a norm against which one may then assess the going practice.

The thesis to be proposed here is that the *basic* or *fundamental* entity in medical practice is the disease; and that the disease is not simply a pathology either of an organ or of an organism, but of the larger context in which the organism grows, survives or dies. If, indeed, the object of medical theory and practice is the treatment of disease, (including here its diagnosis, treatment, cure and prevention) then, I will argue, the very conception of disease (and of its diagnosis, treatment, cure and prevention) varies with the relatively external social and historical contexts of medicine, as well as with the relatively internal development of medical analysis, techniques, and theories of disease. The mediating link between the social and historical contexts of medicine and the "internal" variations and developments of theory and technique is the institutional form of medical practice – including here the hospital, the clinic, the forms of medical education; and the other contexts of individual and collective medical practice, e.g., the life world of the practitioners, his or her social class, the way he or she earns a living, the prevailing value-structure or structures, as well as the relations between medical research and applied practice, between basic research in the sciences and medicine, etc.

In one sense, my proposal is trivial. For if one asks the question, how are doctors *educated* for their profession, the answer clearly is: by learning about diseases. The medical textbook, where it goes beyond the basic information about anatomy and physiology, is a textbook on disease, its diagnosis, prognosis, therapy and cure. Beyond the didactic phase, clinical education assumes an initial acquaintance with differential diagnosis, and presumably teaches the student how to apply what he has learned, and moreover, to learn in the actual clinical situation what cannot be taught from a textbook, or from slides, or in the lecture hall.

But the disease remains, on my view, an abstract syndrome, as long as it remains only a matter of identifying and treating a somatic pathology, say, in a target organ, or even in a whole organism. My argument is that the disease is a social and historical phenomenon, as much as it is an organic pathology; or to put it differently, the organ *and* the organism, in human contexts, have become transformed into social entities, and may no longer be considered abstractly at the biological level alone, without distortion of the very object of medical practice.

In this paper, then, I will, *first*, characterize the human ontology which I would take to be the norm for medical practice; *second*, propose a concept of disease in keeping with this norm; *third*, attempt to relate the important working concepts of organ and organism to this more systematic concept of disease; and *finally*, suggest how the institutional structures of medical practice would need to be adapted to meet this norm.

I do not mean to imply either that there is no such systematic concept of disease already present in some medical practice; nor that present division of labor and specialization are, in themselves, hostile to such a more holistic view. In fact, I will argue that it is the integration of the more specific within the more general contexts that is needed, and not their elimination. The holism I am proposing is therefore an internally differentiated one, in which organ and organism represent levels, rather than merely parts or sub-parts. In this sense, so-called "scientific medicine," incorporating the theoretical and experimental knowledge acquired at the level of cellular and subcellular biochemistry and biophysics, is also incorporated at the appropriate level, as part of the disease-system. In fact, in an optimistic vein, one could argue that present medical practice not only already represents all of these components and levels, but that, in a workaday way, it integrates them as well. But what I am arguing for is a much more deliberately constructed model of the system, *and* a concrete analysis of what it would take to effect it in practice.

First, then, let me pose the question: what is the appropriate characterization of human ontology from which one may derive an isomorphic and appropriate characterization of disease?

It will be useful to eliminate some obviously inadequate views, to begin with:

(a) The human is neither merely a system of cells, nor of tissues, bones and fluids, nor a system of organs. That is to say, neither a cellular, nor an anatomical nor a physiological account succeeds in capturing the reality we call human. That the human is *also* a cellular, anatomical and physiological structure is certainly true. But the very distinctiveness of the human structures, functions and mechanisms at the biological level is itself a function of evolutionary adaptations which are incorporated in the species; and these evolutionary adaptations themselves – from erect posture, dentition, the opposable thumb, the vocal system, the cerebral cortex to the highly specific adaptive immunological system, the digestive tract, and the developmental sequences of physical and psychic growth, – are all marks of a social animal with speech, a culture of artifacts (in the acquisition and preparation of food, clothing, shelter) and a memory which is transposed from one generation to the next by means of these artifacts, and by the keeping of records. In short, the very biology of the human species, whatever correlates it has with other animal species, is a human biology, already infected with culture, history and sociality. Thus, even at the fundamental biological level, species-evolution bears the traces of a distinctive human ontology. One could make the case, holistically, for the effects of this systemic feature of human sociality and culture even at the cellular or immunological levels, but I will leave that as a suggestion only. In short, then, human ontology cannot be reduced to an asocial or ahistorical biology without doing violence to the very specificity of human biological structure and function itself.

(b) Neither is the human to be captured by an abstractive account of psychic life. The idealist tradition in philosophy has long fastened upon thinking, or rationality, or the creation and understanding of meanings as the distinctively human characteristic. There is a profound wisdom in this view; but it has developed one-sidedly, as if psychic life itself were not an emergent rooted in the material and social conditions of human praxis – in the production and reproduction of species-life, in the development of speech and language as a signaling and communicative system, in the social structures which support and ramify human existence.

(c) Nor, again, is the human to be appropriately characterized as a collection of separate and atomic individuals, or persons, fully formed, who then simply unfold their innate characteristics by contracting societies, producing children, and generating history as the liver secretes bile. Such a view – the view typical of so-called theories of civil society

– takes the human individual as only accidentally historical, and social, only after the fact, i.e., by virtue of innate individual capacities or propensities.

If the human is to be appropriately characterized as neither a biological system, nor a rational or meaning-producing entity, nor a collection or aggregate of fully formed individuals, then what is the appropriate view? Clearly, human beings *are* organisms, *are* thinking and reasoning beings, and *are* individuals or persons. The rub comes in abstracting such features from their very genesis in the concrete social and historical contexts in which the species first emerged, in evolutionary terms, and in which it developed its specific historical character, in post-evolutionary or cultural terms (although I would qualify this latter statement by adding that cultural practices – tool production, food acquisition and processing, social structure – already interact with biological contexts in the very evolution of the species). The positive characterization of human ontology would then subordinate the biological, psychic, and personal or individual features to a more fundamental category: the socio-historical and cultural. What this comes to is a realization of individuality, of reason and of organic structure and function as the *products* of an historical development, and as essentially social both in their genesis and in their ongoing modes. Specifically, the individual person is not a genetic prototype which simply unfolds itself in time, but rather the product, at one and the same time, of a "nature" or a genetic inheritance (which is itself an adaptation of bio-culturally evolved species life) and of a "nurture" or a concrete social and historical environment which shapes this inheritance variously. The human understanding, in turn, is itself an adaptively evolved function of social human practice. Further still, the very biological structures and functions are shaped by a life-world which include the distinctively human activities of the production of artifacts, the use of speech, and the evolution of social forms of life which help to determine what is normal, healthy, well adapted to a given historical function, and what is not.

Such a brief sketch of human ontology does not deny either the reality, or the relative separability of individual components or levels of human reality. Cells can be studied as cells, tissues as tissues, organs as organs, individual organisms as individual organisms. That these aspects or components of the human may be identified, even in isolation, *as* human suggests the species-specificity of the parts of the whole, even at the level of the parts. So human blood cells, human bones, and surely human individuals are differentiable by characteristics which may be identified at their own level. To my mind, however, such differentiation already bespeaks the way in which bio-cultural evolution embodies the larger contexts of species-life. In effect, it is the life-world and life-activity – the praxis – of the species which has been mapped into its structures and functions at every level, even down to the subcellular and biochemical levels. Outside of this larger context, the abstracted part or sub-part can be understood only deficiently.

The doctor deals with a patient, but the patient is never an abstract being. He or she is of a certain age, with a certain history, of a certain social class, in a certain social and institutional structure. Thus, the taking of a medical history already connotes the conscious attempt to individuate the patient. But this very individuation is achieved not by isolating the patient, or abstracting him from the range of contexts which define him specifically. Rather, this individuation is a form of concretization, an enrichment of the abstract particular – the mere "this," or "patient *x*" – as a concrete individual, by virtue of a realization of the variety of human contexts which make this individual real. The medical history therefore asks not alone for *medical* information – former diseases, allergies, etc. – but for social and historical information: age, occupation, family history, etc. The fuller history – assigned to the medical social worker – bears as much on intelligent diagnosis and treatment as does the narrower medical information. But that is the point. What the patient is suffering from is a disease, or an injury: and these are social and historical facts about the patient *as much as* they are medical facts. To put it more integratively: the medical facts are *themselves* socio-historical facts.

Having urged that the appropriate human ontology is that of a socio-historical and cultural *system* of individuals, articulated by their concrete and specific forms of life-activity in a given society, I should now map this ontology onto the ontology of a medical practice which recognizes its subject in this way.

What then is the appropriate characterization of disease, in this context? To put the question in terms of ontology, what sort of an *entity* is disease?

Here, again, we may begin by rejecting certain common but inadequate characterizations, which do not capture the reality of human disease as the object of medical practice.

(1) The disease, of whatever sort, may not be adequately characterized as a pathology of an *organ*, isolated from the system in which it functions; nor, at a lower level, may it be characterized, say, as a defect in the bio-chemical mechanisms at the cellular level. Yet, it is obvious that medical discoveries and medical practice *do* focus on organs and on the mechanisms at the cellular and subcellular level, and that some of the greatest advances, both in the practice and theory of medicine, come from scientific and experimental medical research at precisely these levels. But even here, e.g., at the level of immunological research, or at the level of the chemistry of the cell, it is becoming obvious that even in the cell, there are integrative and systemic features which need to be taken into account to understand the micro-processes.

(2) More importantly, and less obviously, I want to argue that the disease cannot be adequately characterized as a pathology of a whole *organism* – i.e., of an individual as a person. Now this sounds perverse, for medical practice *does* indeed deal with individuals and "their" diseases. Yet, I will argue that a disease, or a disease-syndrome, is the property or characteristic of a population, i.e., of a system of individuals in a given socio-historical context. The ontological characterization of a disease is that it ranges over space and through time in a way which transcends its individual occurrences; and, further, that the space-time in which a disease, or a disease-syndrome, exists is a socio-historical and cultural space-time. Thus, it is *not* that the disease manifests itself, or occurs in anything other than individual persons; but rather, that the very characterization of these individuals transcends their particularity.

Let me repeat, in this context, a point made earlier. The individual is a species-being. He or she is *individuated* or concretized – becomes more than an abstract particular – by virtue of the network of relations which determine him or her as this or that individual. Apart from their relations, e.g., living at a certain time, with a certain range of life-activities, in a given culture – in short, as a being socially and historically related to other individuals, to a given environment, and in particular, to a certain historical mode of social production and social life, the individual has no distinctive qualities or properties other than numerical distinctness (being one individual) or self-identity (and these only formally or abstractly).

The matter is somewhat less abstract if we consider the individual as a living organism, for then it shares properties in common with other living organisms. As such an organism, the biological individual constitutes the proper entity for a given level of biological research and experimental practice, but not yet the appropriate entity for medical practice.

The distinctiveness of the human individual, I have argued, rests in his or her sociality and historicity. The fundamental properties of such individuals are therefore *systemic* properties, that is, properties of such individuals not in their isolation, but by virtue of their systemic interrelations, and constituting them as the individuals they are *in* these relations. If this is correct, then disease is to be characterized, isomorphically, as a feature of such a system; or at the very least, of individuals construed in this systemic way. The disease-syndrome, in any given case, is therefore also to be defined as a socio-historical and cultural phenomenon; and the medical practice, whose domain is the treatment of disease, must be conceived accordingly.

Now, let me draw some more concrete consequences of this rather abstract ontological analysis. First, let me present the most plausible cases for such a view. (a) In mass-diseases, medical practice is already constrained to operate within such a context. Thus, with regard to such phenomena as epidemic diseases, communicable diseases, diseases which range over a genetically or ethnically defined population, occupational and regional diseases, and other diseases which can be linked to social class, to characteristic diet deficiences, to specific social or physical environments, and the like, it is clear that medicine deals with disease as a social-historical phenomenon of a given population. Dr. Rosen, in his excellent essay, "Medicine as a Function of Society," [2] makes the case for such a view, in particular, for such classic historical instances as scurvy, pellagra, the "English sweat," syphilis, and others. So, too, Dr. Dubos has characterized a range of mass-diseases in socio-historical context, as the "Diseases of Civilization." [3]

(b) At another level, diseases specifically related to such apparently biological characteristics as age or sex also range over differential populations. Pediatric, gynecological, and gerontological diseases are therefore also at least systemic, rather than merely individual, considered as objects of medical practice. But even these apparently biological characteristics are in fact conditioned by socio-historical context. The concepts of infancy, adolescence, middle age, and old age, for example, are cultural artifacts, and have emerged in different ways in different societies. And it is clear that such "artifacts" have affected medical practice, and have often been affected, if not constituted, by medical practice.

(c) What remains is the isolable somatic disease or the specific or local injury, which escapes the earlier characterization. But which are these? Renal diseases, or endocrine malfunctions, or circulatory and respiratory diseases, or diseases of the digestive tract, or mental diseases – all, in short, may be seen to exist in the richer contexts of particular social environments, particular historical periods, particular life-forms and life-activities. So, too, they may be seen to exist, in a different sense, by virtue of the very structures, institutional forms and practices of medicine itself. I am not saying that medical practice *creates* these diseases (though there are such instances as well, childbed fever being a classic case, and the side effects of drug and other therapies being another class of such cases), but rather, that a given form of medical practice – including here its methodology, its institutional forms, etc., – picks out certain syndromes as typical or characteristic ones in its repertoire. We may therefore conceive of diseases which already exist in a population but have not yet been discovered, or differentiated, or which have been confused or misidentified with known diseases. In this sense, then, there is a range of diseases which are artifacts of medical practice in these two senses: (a) they are the products of medical practice, i.e., are caused by certain practices; (b) they are discovered, isolated or differentiated by medical practice. (The first case is malignant; the second, benign.)

Now, these remaining diseases may therefore be seen to be systemic, or features of a population, and of socio-historical contexts in all of these cases. But is this in effect a trivial thesis? For if we define *any* disease – e.g., kidney disease – as a disease shared by all those who have it (including the historical set of all who have had it in the past and who will have it in the future) haven't we defined the "population" or the system in a vacuous way, namely, as that population which has the disease? If the alleged "socio-historical group" which defines kidney disease is nothing more than the set of all those who have the disease, then the notion of a system or a population is indeed vacuous and the definition of the disease-group tautological.

It is not vacuous from another point of view, however. Comparative studies of all those who suffer from a given disease may indeed reveal common features in the etiology of the disease, in its course, or in its systematic relation to other factors which may in turn lead to non-vacuous discoveries which bear on the understanding and treatment of the disease. Any number of levels may be relevant here: the isolation of biochemical causal agents, of environmental or genetic causes or conditions, or indeed, of social and cultural factors, may result from such comparative studies of a disease-population.

But such diachronic or synchronic studies of a disease population are precisely the ways in which the apparently vacuous abstraction – e.g., "all those with renal diseases" – becomes concretized and, I would argue, socialized and historicized as well. Even in the case where the social environment or social forms of life-activity seem least relevant – e.g., in genetic diseases – population genetics (where it is not simply an abstract statistical or mathematical discipline) already intrudes such considerations as demography, population-migration, forms of kinship and family relation as they bear on the reproductive patterns, etc.; and these, in turn, surely involve social, historical, and cultural considerations.

In summary, then, all such diseases *ought* to be conceived, in their broader contexts, in parallel with the human ontology which is enmeshed in the sociality and historicity of human beings, in all of their functions.

The trouble with the foregoing is that it seems to ask the doctor to stop treating the organ or the organism, and to start treating the population as a whole. And populations are notoriously abstract entities to treat, in the ordinary practical contexts of medical practice. Perhaps the epidemiologist, the medical statistician, the medical sociologist, the historian of medicine can deal with such abstract conceptual entities as a "population" – the practitioner, it may be argued, deals with Mrs. Jones, Mr. Kowalski, and little Mary Ann; and moreover, with Mrs. Jones's sciatica, Mr. Kowalski's lung cancer, and little Mary Ann's fractured tibia. It would also seem that my proposal would require the practitioner to become a generalist of such proportions that he could deal only in the vaguest generalities and would never learn the hard details of his or her craft which are necessary to pursue it well.

Let me say two things about that.

First, the medical practitioner doesn't deal with an individual patient, merely, no more than the college professor engages in one-to-one teaching, or fulfills the utopian image of the professor at one end of a log and the student at the other. The doctor meets the patient within a social and institutional structure: in private practice, the clinic, the emergency ward, the big city hospital, the community health center, and even occasionally in the patient's home. In these institutional settings, even such socio-cultural artifacts as the available pharmacopia, the instrumentation available, Blue-Cross-Blue-Shield, Medicaid, Medicare, and the guild ideologies and practices of the A.M.A. all make a difference, both in modes of diagnosis and treatment.

True, medicine, like justice, should be blind to persons, and deal equally with all: as the punishment should fit the crime, so should the diagnosis and therapy fit the disease, regardless of persons. But this is, in the practical world, as much a myth in medical practice as it is in legal practice, though it is an ideal myth. Still, Mrs. Jones, Mr. Kowalski, and little Mary Ann *are* treated in specific settings, and bring specific expectations, fears, and capacities to respond to treatment, because of their social and historical realities. In the same way, the doctor brings his conceptual framework, his specific and even idiosyncratic modes of practice, his ideology, and his professional specializations and limitations to bear on his actual diagnosis and therapy. In this sense, the most *individual* medical care is already caught up in the network of social relations which constitute the medical practice of an age, or of a given society.

Second, I am *not* arguing for an abandonment of a high degree of specialization, or of the most thorough training in basic medical sciences and in the specifics of anatomy, physiology, pathology, etc. The advance and revolutionization of medical practice has proceeded, as in other human endeavors, by division of labor. But even in the most clearly separable cases, e.g., those of specific injuries, or localized infections, I want to argue that the larger context does come into play, and that to ignore it, in our general conception of medical practice, is to fall prey to a fragmentation of medicine itself which will ultimately prevent it from dealing adequately with the very entity which defines its practice – i.e., disease.

In short, to state the matter in a somewhat caricatural fashion, the practitioner dealing with a localized infection of the tissue of the left leg is, at the same time, caught up in a network of practices, attitudes, and institutional forms which in some sense determine the modes of his specific treatment. The reason for insisting so strongly in this way is to counter a common mystification, at least in the popular conception of medicine, if not also one which helps to shape medical practice and medical education: namely, that, whatever else may or may not be relevant, when one is treating the leg, one is treating the leg, and nothing else matters. My point is that in treating the leg as the leg, everything else in fact *does* matter, and affects the ways in which this very isolable and individual case is dealt with. I am not arguing *against* the isolation and focus on a given highly specific organ or tissue and its pathology, but rather that this relative isolation and focus is a *tactic* of medical practice, and neither an appropriate strategy nor a definition of the real subject of treatment. If this isolation were more than a tactic, then medical practice itself would become fragmented and atomized, and medicine would become an agglomeration of discrete and isolated specialties, whose fundamental ontology would be atomistic: organ by organ, tissue by tissue, with the diseases themselves seen as discrete and unconnected syndromes, at best.

Since we all *know* this *cannot* be the case (or at least, that where it is the case, it is wrong), one should ask what would prevent (or cure) such a pathology of medical practice. Since specialization in medicine is a *sine qua non* of competence and adequacy in the diagnosis and treatment of specific diseases, to argue *either* that specialization alone is the *cause* of such fragmentation, or that one ought to do away with it is to become a medical Luddite: "Smash the machines (or the specialities) and go back to 'nature' (or the whole organism) and we'll all be fine."

Thus, it is not specialization as such, but rather the conceptual, institutional, and practical forms which such specialization may take (or has taken) which result in an atomized medical ontology and an atomized medical practice. But if the "whole" entity for medicine is not even the organism as a whole, the human being, but the *disease*, then the appropriate ontology is one which reflects the full richness of interconnections which a disease syndrome presents: not an abstract and seamless "whole," but rather an intricate and many-layered network of social, personal and organic contexts – from society to cell, so to speak – in which the doctor intervenes at specific points, diagnostically and therapeutically, because the whole system is understood in a certain way.

In conclusion, let me briefly state what I think would be an appropriate strategy for the integration of the levels of organs and organisms within the larger construct, *disease*, as I have characterized it. Then, finally, let me suggest what therapy is needed for the pathology which threatens medical practice with fragmentation and disintegration. In both cases, the individual practitioner will be able to achieve this integration not as an isolated individual, but only as a member of a collective, guided by an ideology or a heuristic principle which is in accord with or appropriate to the function and aim of medical practice. We shall see that, reflexively, what I have said about human ontology and the parallel ontology of the object or fundamental entity of medical practice bears as well on the organization and form of this practice itself, i.e., on the methodology of medicine.

(1) First, the forms which mediate the social-historical life-world of a population, its concomitant illness and health on the one hand, and the medical practice which deals with this human world on the other, are, as I have said, the institutional forms of medical practice, including medical education. The resolution of the problem lies within those institutional forms and structures – within their present potentialities, and the changes which may be necessary. Thus, it is not enough to conceptualize the integration of which I am speaking, as if, once resolved in the *minds* of physicians, it would then be effected in their practice. The need for specialization militates against that. It does help, however, for the individual physician to be concerned, in general, with the integration of his profession. Insofar as he or she participates in determining its practice and in molding its institutional forms, his or her general views of social and medical reality – what one might call his or her ontology, or even his or her metaphysics of medicine – play a role heuristically, and also politically. But such conceptual or epistemological reform is not enough. It has to be embodied in the actual structures and practices of medicine, in their institutional forms.

How then is the treatment of a localized infection or the disease of a specific target-organ to be integrated into the larger context? By the socialization of the forms of medical practice. I am not talking about socialized medicine in the ordinary sense, of the social responsibility for, and the social funding and organization of, health care, although I do believe that this is one of the essential social conditions for the restructuring of medical practice itself. That is, it is my view, in general, that a fundamental change in society, and in social priorities, remains one of the conditions for effectively changing medical practice in a radical way, and that such a socialization and humanization of medicine cannot be brought about merely internally. But however *such* socialization of the availability and delivery of health care is effected, I am talking here in a more limited way about the socialization of the internal *practice* of medicine, a socialization which has already proceeded quite far.

Let me give some concrete examples. The surgical team is already an example of a socialized practice. Here, the division of labor and specialization is subordinated to an integrated function, surgery, in which the interrelation of the various subfunctions is clearly and efficiently organized to serve a given end. So, too, is the method of consultation and evaluation. And, in the optimal hospital setting, the doctor functions within the context of a complex set of support systems, all of which are subordinated and "socialized" towards a given end: the pathology laboratory, the nursing staff, the x-ray laboratory, the pharmacy, the attendants, volunteers, the orderlies, the heating maintenance staff, the medical social worker, and optimally the patient's family as well, are focused on the given case. With the development of health centers, medical research and education are also integrated, for better or worse, in the system; and with computer terminals, telephone and television communication across long distances, the "center" becomes a regional or even national or international network.

Such a "socialization" of medical practice already exists, therefore, in some measure. The problem is to realize it, not by accident, nor by the occasional confluence of favorable conditions, but by conscious design; and not to effect merely a conglomeration of unintegrated services, each with its guild-barriers high, and its historically uneven development (e.g., the unenlightened and dysfunctional role to which nurses are assigned, by virtue of a defunct and anomalous tradition, and the depersonalization of the patient to suit imaginary "hospital" needs which date back to another era, etc., etc.).

In short, unintegrated specialization leads to fragmentation where the teleology of individual practice is poorly understood, where the ends of medicine are themselves fragmented, and where dominance and subordination with respect to a given end reflect social and historical patterns of domination inherited from the past, or reflect social class and occupational hierarchies in a society, which are *not* relevant to medical practice or its ends as construed above.

These considerations bear directly on how one is to treat the organ, and the organism, in this larger context. The issue is not simply one of the epistemology and ontology of medicine, but also of the consequences in practice of one or another epistemological and ontological view.

(2) It is not the case that the doctor is inadequate to this task, or that he has to give up his specialization to achieve integration. Rather, he has to construct his professional life in such a way that he becomes a collaborator in the system of health-care. This means, in part, an awareness in principle of the available medical technology, the other specialties, etc. – i.e., a functional-systematic map of the whole enterprise. This does not mean expertise in many fields, but a liberal education in medicine itself, to overcome the parochialism which is induced by division of labor (as it normally proceeds in medical education and professional life). Now there may be a need for specialists in integration – a function which, for example, the internist serves at one level, the G.P. at another. What may in fact be required is the development of the medical generalist as a *high* specialist. The hospital administrator is (or ought to be) just such an organizational specialist – i.e., in the organization of services. But his equivalent in functioning medical practice is still needed.

But beyond this, the practitioner needs not only to *know* the integrative norms required for an adequate medical practice. He needs also to be a political person in effecting such changes in the present forms of medical advancement and medical practice. This requires a critique of existing institutional forms and medical praxis, from within medicine itself, so that the social and political changes in medicine and in health generally are not simply imposed from without, out of radical social needs, without due regard for the medical parameters themselves.

Thus, the system of disease, as a socio-historical system, demands not only a normative ontology of practice, but a reconstruction of the forms of that practice itself, in order to become more adequate to the reality of ill health and good health, which is the object of that practice. Human ontology thus becomes a model – a heuristic and normative model not alone for the reconception of disease, but for the forms and institutions of medical practice itself.

Boston University,
Boston, Massachusetts

NOTES

[1] René Dubos, "The Diseases of Civilization," in *Mainstreams of Medicine*, ed. by Lester S. King (Austin and London: University of Texas Press, 1971), p. 52.
[2] George Rosen, "Medicine as a Function of Society," *Mainstreams of Medicine*, ed. by Lester S. King (Austin and London: University of Texas Press, 1971), pp. 26–38.
[3] Dubos, *op. cit.*

PATRICK A. HEELAN

COMMENTS ON "CONCEPTS OF FUNCTION
AND MECHANISM IN MEDICINE AND MEDICAL
SCIENCE" AND "ORGANS, ORGANISMS
AND DISEASE"

The first great work on the philosophy of medicine is Claude Bernard's
An Introduction to the Study of Experimental Medicine[1] which was written
just over one hundred years ago. Bernard was one of the first proponents
of an organismic vision of life – neither crudely mechanistic nor vitalistic
– that affirmed both the universal validity of physio-chemical laws in the
biological domain and the existence of special non-reducible features,
physiological in character, of living matter. For Bernard, the body was
a "living *machine*"[2] not exempted from the laws of physics and chemistry,
and a "creative *idea*" which "expresses itself" through physio-chemical
means. Physio-chemical means, he continues, "are common to all natural
phenomena and remain mingled, pell-mell, like the letters of the alphabet
in a box till a force goes to fetch them, to express the most varied thoughts
and mechanisms."[3] For Bernard, then, organs were like *words*, and individual organisms were like *sentences*. Here Bernard's – and Toulmin's –
analysis would stop. Wartofsky, however, would go one step further and
say that socio-historical communities of organisms are like *languages*:
diseases which for Bernard and Toulmin then are pathologies of "words;"
and "sentences" are for Wartofsky principally pathologies of "languages."

Professor Toulmin's paper is concerned with the inseparability and
irreducibility of the basic notions of *mechanism* (based on *physics and
chemistry*) and *creative idea* (or *purpose* or *function*) in physiology. I am
in fundamental agreement with Toulmin's conclusions. I hope it is not
presumptuous for a commentator to add something to Toulmin's ideas
by showing how it is possible to formalize in a context logic the relationships he establishes between the language, L_m, of mechanism and the
language, L_f, of function.

Professor Toulmin has shown that *mechanism* and *function* are partial
descriptive aspects of one real, organic system that is within one context
of inquiry a mechanism neutral to values, within another context of inquiry a function laden with values, and in yet a third context both a
mechanism and a function simultaneously.

H. T. Engelhardt, Jr. and S. F. Spicker (eds.), Evaluation and Explanation in the Biomedical Sciences, 85–93.
All Rights Reserved. Copyright © 1975 by D. Reidel Publishing Company, Dordrecht-Holland.

In terms of context logic[4] – the logic of relations between languages based on their respective semantical power[5] – we have the following state of affairs (see Figures 1 and 2).

$L_{m \oplus f}$ is the combined language of mechanism and function. The set

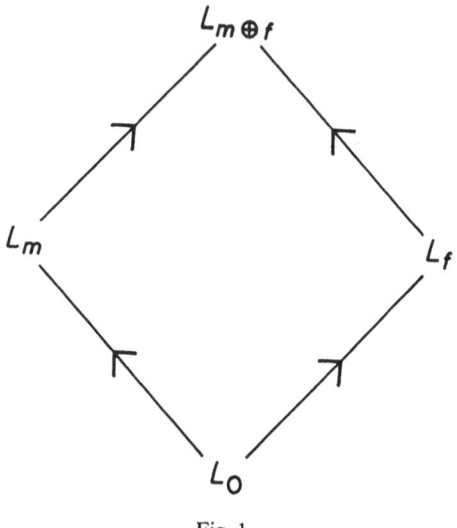

Fig. 1.

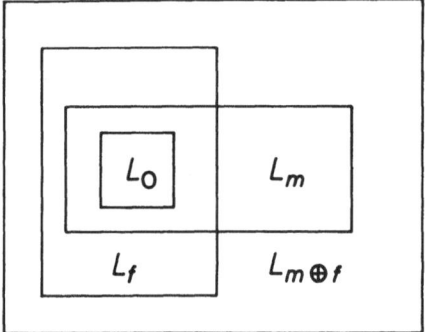

Fig. 2.

$\{L_0, L_m, L_f, L_{m \oplus f}\}$ constitutes a lattice under the partial ordering "$\rightarrow$—" ("implies"), which has the following interpretation "$L_i \rightarrow$— L_j (L_i implies L_j) if and only if whatever can be said in L_i can be said in L_j but not necessarily vice versa." In semantical terms, L_j is a *conservative extension* of L_i.[6] The lattice, moreover, is complemented as will be shown. When the complements are added, we get the following representation (see Figures 3 and 4), where L'_m and L'_f are complements respectively of L_f and L_m. The lattice is non-distributive, since,

$$L'_m \otimes (L_f \oplus L_m) = L'_m \neq L_m = (L'_m \otimes L_f) \oplus (L'_m \otimes L_m)$$

provided $L_m \neq L'_m$. Likewise, provided $L'_f \neq L_f$, we have

$$L'_f \otimes (L_m \oplus L_f) \neq (L'_f \otimes L_m) \oplus (L'_f \otimes L_f).$$

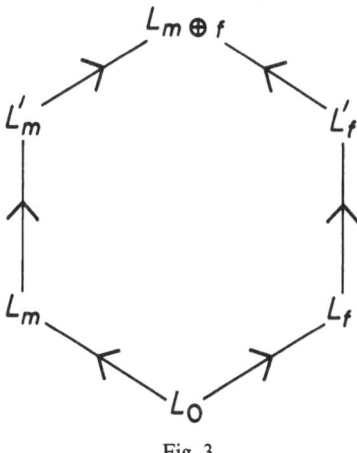

Fig. 3.

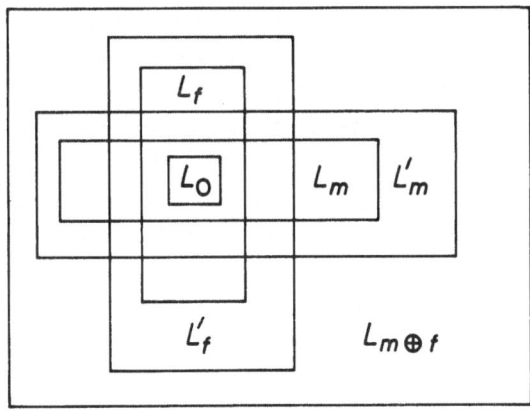

Fig. 4.

Hence, products (g.l.b. [greatest lower bound] or $\otimes$) do not distribute over sums (l.u.b. [least upper bound] or $\oplus$). L'_m is a conservative extension of L_m in which, in addition to the vocabulary of physics and chemistry, functional terms are also used, but in a non-functional, mechanistically reductionist sense, following, for example, Nagel's reconstruction of teleological language. For Nagel, to every teleological account there is a semantically equivalent non-teleological translation[7]: these semantically equivalent non-teleological translations belong to L'_m. I take it that Toulmin and Bernard are saying that the teleological language L_f is *not* equivalent to L'_m. If Nagel's position is free of contradiction – and I suppose it is – then L'_m, the reductionist reconstruction of teleological language, exists – even though, in our view, it may not be semantically equivalent to L_f.

The existence of L'_f is harder to show: it would bear a resemblance to the language of Aristotle's *Physics* (or Aquinas's cosmological philosophy) in which the causality of mechanical events is interpreted in reductively teleological terms.

It follows then that the set $\{L_0, L_m, L_f, L'_m, L'_f, L_{m \oplus f}\}$ exists, and that the elements are related by the lattice relationships depicted in Figure 3 and the extensional relationships depicted in Figure 4 (the areas represent the semantical power of the language).

The significance of what I have just proven – that the languages L_m and L_f of mechanistic and functional description respectively belong to a quantum logical lattice (complemented and non-distributive) – is that it shows the kind of rationality associated with the use of mutually irreducible frameworks like L_m and L_f, which are also complementary in the sense made familiar by quantum mechanics. (Note the two different senses of the term "complement" or "complementary," one strictly mathematical, the other belonging to the Copenhagen Interpretation of quantum mechanics.) The lattice of Figure 3 depicts the context-logical analysis of quantum mechanics – where L_m and L_f are two complementary frameworks in the quantum mechanical sense. The lattice is called a "quantum logic" – a non-distributive complemented lattice. The dualism of mechanism and function, then, is logically parallel – in context logic – to the dualism of complementary frameworks, like precise position and precise momentum, in quantum mechanics. There is even a parallel uncertainty principle: the more exact the specification of a particular physio-chemical pathway, the less likely it is that its function will be exactly specified, and vice versa; for good functioning is a property manifested only over many cycles and usually over many different physio-chemical pathways, while the same physio-chemical pathway can be travelled by many different functional messages.

What I have said about a quantum mechanical model for the complementary languages of mechanism and function in biology and physiology has been said before; Prof. Toulmin has in fact alluded to this view in his paper. But since quantum mechanics is an obscure science, surrounded at all times since its birth with controversy, there is the danger of explaining the obscure by the more obscure. It is my contention that quantum mechanics is the first science to incorporate within its semantical model not merely a logic of sentences, but an intrinsic logic of irreducible languages. This structure has been elucidated in the kind of quantum logic I have called "context logic,"[8] and which I used above to express the relationships between the elements of the set of languages $\{L_0, L_m, L_f, L'_m, L'_f, L_{m \oplus f}\}$.

The notion of *level* – notoriously ambiguous because relying on a metaphor – can also be explicated through context logic, for the partial ordering of implication ("$\rightarrow$") traces out a hierarchical ascent on condition that any subset of languages descriptive of partial aspects of the organism, have a l.u.b. (least upper bound), that is, on condition that the context logic is a lattice. Thus, an organism which integrates two and only two systems would fulfill the following diagram (see Figure 5), where L_1 and L_2 are the sub-lattices of the two component systems and $L_{1 \oplus 2}$ is the complete descriptive language of the whole organism. There are six levels identified in Figure 5.

Since it is the usual function of a commentator to raise questions for a future discussion, let me briefly do just that. Prof. Toulmin defines a *function* as a process that provides the *right output* (of information in some physio-chemical channel) for given *occasions* (of input information) in pre-determined *loci*. The process *uses* physio-chemical mechanisms to achieve the desired results. To the extent the process is guided by a norm of *right* operation, e.g., to stabilize the temperature of the blood or what have you, the process is a *function*: a function then incorporates a *value* – the value of good functioning or health for the organism. I want to point out, however, that health or the good functioning of organs and systems is a value *for the organism*: value-notions, however – not values – are what are constitutive of the scientific understanding of organic functions. Hence, one cannot argue that the health of the organism studied is necessarily a value for the scientist or physician studying the organism, though the notion of health is. The commitment to safeguarding the

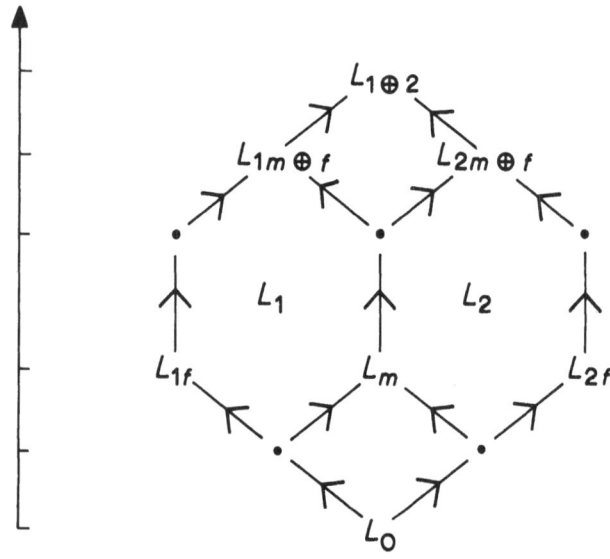

Fig. 5.

health of the organism he is studying is – contrary to what Prof. Toulmin seems to hold – an ethical commitment *over and above* the use of value-laden notions in explaining organic functions. In this, I appear to differ from Prof. Toulmin, whose statements, however, I find confusing. He says: "The chief vital functions of the human body are not merely 'good in themselves.' They are preconditions for almost any other imaginable human good."[9] But what does Prof. Toulmin mean by "good in themselves"? Does he mean a good *for all*; for *n'importe qui*, and therefore to be respected and pursued by *n'importe qui*? Or does he mean a good for the organism, the truth of which is to be acknowledged by *n'importe qui*? The sequel in the text seems to say that the health of the organism is a good for all: "It is no longer the case that the physician has nowadays to make any specifically ethical commitment to sustain health":[10] he has implicitly acknowledged, "taken it for granted," "made the assumption" that the health of the organism is both "desirable" as well as "actually desired."[11] However, the reason he gives is that the physician's "whole enterprise is nowadays intelligible, and capable of being rationally expounded, only in terms of a systemic picture of vital organization and functioning."[12] But systemic intelligibility (for the physician) based on the notion *value for the organism* surely does not imply that the physician by virtue of the assumptions of his knowledge, acts to preserve that value for the organism. This would follow only if there were a specific ethical commitment on the part of the physician to identify his own good in action with the good of the organism he is studying or treating. To deny the need for a specific ethical commitment would be to conclude the logical impossibility of forwarding scientific knowledge by vivisection where the animal dies as a consequence of the experiment. Claude Bernard, the great proponent of vivisection as necessary to experimental medicine, would never have agreed with this conclusion, nor do I. I feel, therefore, that Prof. Toulmin has either not expressed himself clearly or tried to be too subtle and fallen into a variant of the Platonic trap of identifying knowledge of virtue with the possession of virtue.

Among his many credits, Prof. Toulmin was a pioneer in the study of "scientific revolutions." To know about revolutions, however, is one thing; but to institutionalize it as a philosophical value – as the essence of true philosophy – is another. Wartofsky would do the latter. Philosophy for him is *dialectical* in the sense that beyond theories, conceptual models, common sense or traditional maxims is the *praxis* of the sociohistorical milieu out of which comes the critique of old forms of knowing and the emergence of new forms. The old forms of knowing – whether expressed in theories, models, common sense or traditional maxims – become in the moment of their expression abstract, static and restrictive, co-opted in the service of manipulation and control, and impositions of static mind on fluid process. They have lost their power to manifest and celebrate being. There is a way – a very significant way, Wartofsky would say – in which one thinks with one's body – with one's hands and fingers, and with tools and instruments that encounter reality and shape our concept of reality while shaping reality itself. Out of such encounters come theories, models, common sense and traditional maxims, only to be criticized and replaced in turn as thought, the slower runner, catches up with the movement of praxis.

Such – if allowance is made for an element of caricature – is Wartofsky's critique of essentialisms, conceptualisms, and idealisms of all kinds. It is, I believe, a critique of the utmost importance, given how prone we are – in this scientific community – to capitulate uncritically to the authority of theories and to the authority of specialized information and advice. By *uncritically*, I mean failing to pass such knowledge through the critical mediation of social and historical praxis, or not knowing how to perform such a mediation.

However, in appealing to the broadest context of social and historical praxis as the appropriate context within which to locate the entity of disease, Wartofsky may have left us confused and resourceless with respect to the present treatment of disease. If a disease is a socio-historical entity, then treatment or therapy *ought* to look not merely towards ameliorating the present situation, but even more perhaps towards the future shape of the socio-historical context: therapy must then take on an eschatological dimension, become conscious of the shape of things to come. We become dependent on prophecy, or on a not-yet existing science of the future. Thus, we are left resourceless and confused. By insisting that the notion of *disease* be defined only in the largest context, Wartofsky may have deprived us of the *long-term* means of treating disease. Moreover, since *disease*, as Wartofsky tells us, is to be defined as correlative to *treatment*, not having clear knowledge of *long-term* treatment deprives us of clear knowledge about *long-term* disease. On the other hand, we do know a great deal about both diseases and their treatment *in the short term*. What then are physicians? Are they healers of the *long-term* somatic ills of individuals in society or merely of their *short-term* somatic ills? Asked to choose, I would personally choose a physician skilled in short-term treatments and believe that such is what a physician should be; although the medical profession, as a whole, and medical practice, through its insertion into a socio-historical community, should be responsive to social goals formulated within the wider community.

State University of New York at Stony Brook,
Stony Brook, New York

NOTES

[1] Claude Bernard, *An Introduction to the Study of Experimental Medicine*, trans. by H. C. Greene (Dover: New York, 1957).

[2] *Ibid.*, p. 89 (italics added).

[3] *Ibid.*, pp. 93–4.

[4] Patrick Heelan, "Complementarity, Context-dependence and Quantum Logic," *Foundations of Physics* 1 (1970), 95–110.

Patrick Heelan, "Quantum Logic and Classical Logic," *Synthese* 22 (1970), 3–33.

Patrick Heelan, "The Logic of Framework Transpositions," *International Philosophical Quarterly* 11 (1971), 314–334.

Patrick Heelan, "Hermeneutic of Experimental Science in the Context of the Life-World," *Philosophia Mathematica* 9 (1972), 101–144.

[5] By "semantical power," I mean the expressive power of the language, which is, roughly, the measure of the set of specifically different elementary situations describable using the resources of the language in a literal fashion: elementary situations are mapped on elementary categorial statements in the language.

[6] P. M. Williams, "On the Logical Relations between Expressions of Different Theories," *British Journal of Philosophy and Science* 24 (1973), 357–67.

[7] Ernest Nagel, *The Structure of Science* (London: Routledge and Kegan Paul, 1961), pp. 401–28.

[8] P. Heelan, "Complementarity, etc."

[9] Stephen Toulmin, "Concepts of Function and Mechanism in Medicine and Medical Science (Hommage à Claude Bernard)," in this volume, p. 61.

[10] *Ibid.*

[11] *Ibid.*

[12] *Ibid.*

SECTION III

ETHICS AND MEDICINE

ALASDAIR MACINTYRE

HOW VIRTUES BECOME VICES: VALUES, MEDICINE AND SOCIAL CONTEXT

I

I begin with three distinct groups of problems, each so urgent that the medical conscience ought to be – and indeed clearly is – haunted by them – and not only the medical conscience. Consider first such problems as: ought abortion to be legal? Is it ever morally right? Ought those in extremes of pain to have the right to take their own lives? Ought physicians to have the right to take the lives of patients with terminal cancer, who are in extreme pain?

Or consider secondly such other problems as: Should a physician be obliged to tell a patient that he is dying? Should a physician be obliged to give a patient a true diagnosis about his condition? Should a hospital be obliged to publish each year the percentage of operations in which after an organ was removed it was found to be undiseased and undamaged?

Or consider thirdly such rather different problems as: What is the justification for paying such high salaries to American physicians and surgeons as compared with the salaries of their counterparts in certain other countries, when in infantile mortality rates and the life expectancy of a man aged 45 the United States lags substantially behind those countries? Ought the supply of medical care to be surrendered to the demands of a free market economy? Ought the sole criterion for the availability of medical care to be need? How should resources be allocated between different needs?

The first set of problems concerns the relationship of medical practice to the good of preserving human life; the second concerns the relationship of that practice to the goods of trust and truthfulness; the third concerns the relationship of medical practice to the good of justice. Each group of problems is notoriously a matter for contemporary moral debate. I am going to argue that we have no rational method available for reaching a conclusion on these questions and that this springs from the

H. T. Engelhardt, Jr. and S. F. Spicker (eds.), Evaluation and Explanation in the Biomedical Sciences, 97–111.

peculiar character of moral debate in our liberal, secular, pluralist culture, rather than from the special character of these problems or from the general character of moral argument.

II

One way to approach this thesis is to set out examples of opposing arguments on one out of each group of issues:

1(a).   I cannot will that my mother should have had an abortion when she was pregnant with me, except perhaps if it had been certain that the embryo was dead or gravely damaged. But if I cannot will this in my own case, how can I consistently deny to others the right to life that I claim for myself? I would break the so-called Golden Rule, unless I denied that a mother had in general a right to an abortion.

1(b).   Everybody has certain rights over his or her own body. To establish such rights we need merely to show that it cannot be shown that anyone else has a right to interfere with the implementation of our own desires about our bodies. It follows that at the stage when the embryo is essentially part of the mother's body, the mother has a right to make her own uncoerced decision on whether she will have an abortion or not. Since she has a moral right, she ought also to have a legal right.

2(a).   A physician should decide whether to tell a patient who is gravely ill or dying the truth about his or her condition by reckoning on the consequences in that particular case to that particular patient of giving or withholding that specific information. If the patient's health and happiness will be increased by telling the truth, then the truth should be told; but if not, not so.

2(b).   To treat an agent with moral respect is to look to his dignity and not his happiness. To deprive a man of the truth about his disease or his death is to deprive him of dignity. More particularly every man can only respect himself if he faces up to the fact of his own death. Hence to deprive patients of the truth about themselves is to do them a wrong and a wrong that insults their status as human beings.

3(a). Justice demands that every citizen should have so far as is possible an equal chance to develop his talents and his other potentialities. But good health is a prerequisite for such development. Therefore every citizen should have an equal right to access to the means of good health, so far as it is available. Therefore justice requires a free national health service, financed out of taxation, with no private sector of medical practice.

3(b). Everybody has a right to incur only such obligations as he chooses, to be free to make such contracts as he wishes. Every doctor therefore must be free to accept patients or not and to practice on what terms he chooses. If others then do not wish to deal with him, that is their free choice. Freedom requires freedom of contract. Freedom of contract requires medicine to be a matter of private practice.

About these six arguments I want to make two main points. The first is that the goods which each argument aspires to vindicate are incommensurable with the goods invoked by at least some of the other arguments. Each of the arguments moves validly from premises to conclusion; but the premises are so independent of each other that those who accept the premises of rival arguments share no common moral ground. Given each set of premises, it is rational to move to each conclusion, but there is no argument available, no criterion available, no rational procedure to decide between rival and incompatible conclusions.

We appear therefore to have to make a non-rational choice between alternative positions, so far as our own moral judgments upon each issue are concerned, and to have to resort to non-rational persuasion if we are to affect the choices made by others. What puts us in this position? One standard type of answer is that it springs from the nature of morality as such. C. L. Stevenson's position in *Ethics and Language* requires that moral disagreement be rationally interminable.[1] Sartre's moral philosophy – at least in one version – notoriously makes an arbitrary choice the fulcrum upon which moral judgment depends. If they and those who think with them on this matter are correct, then the place of arbitrary choice and of interminable disagreement in morals will be the same irrespective of time and place, irrespective of the differences between different social and cultural orders. But is this true? As a preliminary to raising and trying to answer this question, it is important to glance at the historical background to the present situation.

We ought first to note that the historical provenance of the six arguments which I outlined is very various. So far as their philosophical ancestry is concerned, the first has its roots immediately in Kantian thought and more remotely in Christianity, the second in the thought of Jefferson, Robespierre and Rousseau, the third in Benthamite utilitarianism, the fourth in Hegel and Kant, the fifth in T. H. Green and in Rousseau, the sixth in Adam Smith and John Locke. But of course the premises of the six arguments have a genealogy which extends beyond explicitly philosophical texts, indeed far beyond any texts. For some of these premises at least have informed a variety of types of social and moral practice and as such have defined in part cultures which are our moral and intellectual ancestors. But then they were parts of wholes, ordered within some total moral scheme which provided a vision of man's true end, of the relation of his empirical nature to his essential nature. It is a tacit assumption of secular, liberal, pluralist culture, of the culture of modernity, that to a rational man no such vision is now available, because we can have no rationally defensible concept of man's true end or of an essential human nature.

Consequently what we inherit from the varied and different strands of our past is a collection of fragments, of moral premises detached from the contexts in which they were once intelligibly at home, survivals now available for independent moral assertion from a variety of moral points of view. It is this that makes moral argument appear to consist merely of the clash of bare assertion and counter-assertion marked by what is only the appearance of argument, so that non-rational persuasion seems to be the only way to terminate disagreement, and arbitrary choice seems to be the only way for an agent to resolve the issues in his own mind.

That is to say, the Stevensonian or the Sartrian conclusion expresses a genuine sociological insight into the fate of morality in modern culture, rather than a philosophical insight into the nature of morality as such.

But the Stevensonian or the Sartrian – or a Nietzschean – philosophical claim about morality could only be vindicated if no thesis about essential human nature and its true end could be made good.

Does this mean that in the present historical situation we are doomed to arbitrary non-rational solutions to moral problems, including the moral problems of the medical practice? In order to answer this question we must first enquire how the general character of the moral history of our culture has affected the special character of medical ethics. For it might be claimed, and it often has been claimed, that the morality of the medical profession is in a special way autonomous. The medical profession has had to safeguard and to transmit its values in a variety of very different social contexts: late medieval Catholicism, renaissance Italy, eighteenth century France, twentieth century America. The very claim to be a *profession* and not merely a trade characteristically involves a claim to be the guardian of certain values.

The values to which the medical profession has specifically claimed to be committed through changing social and moral environments are at least threefold: there is the unconditional commitment to preserve life and health, the responsibility for justifying the patient's trust in his physician or surgeon, and the demand for the autonomy both of the individual physician or surgeon in making his judgments and of the profession as a whole in sitting in judgment on its own members. The crucial question is: could these virtues not only survive in their integrity, but provide an adequate moral criterion for medical practice in a culture with a history of moral fragmentation such as ours?

At first sight it might seem at least not impossible that this should be so. For certain virtues incarnate goods, the rational claim of which upon mankind seems to be independent of cultural history and variation. In order to understand why this can be so, I must give a more general account of the virtues.

III

It is a commonplace that societies differ in their accounts of the virtues. There are not only differences as to which human qualities are to be accounted virtues, but there are of course also differences as to the criteria by means of which the virtues are to be justified. Fifth century Athens is in many ways at odds with twelfth century Iceland; Polynesia is radically different from Pennsylvania. But it is a condition of our ability to point out this truth (so often one of the premises from which a facile relativism is deduced), that we are able to identify in each of these societies one and the same focus upon a set of human qualities, the presence or absence of which in a man determines how he is to be assessed as a man. Were this not so, we would not be able to understand these various cultures as differing from each other over one and the same thing. Our perception of difference presupposes a perception of resemblance.

What then makes a virtue a virtue? We define and we cannot but define our relationships to other people by referring to certain goods. A, B, C and D are friends. D dies in obscure circumstances, A discovers how D died and tells the truth about it to B, while lying to C. C discovers the lie. What A cannot then intelligibly claim is that he stands in the same relationship of friendship to both B and C. By telling the truth to one and lying to the other he has partially defined a difference in the relationship. Of course it is open to A to explain this difference in a number of ways; perhaps he was trying to spare C pain or perhaps he was ashamed to reveal the nature of his enquiries about D to C or perhaps he is simply cheating C. But some difference in the relationship now exists as a result of the lie.

Just as we define our relationships to each other, whether we will it or not, by reference to standards of truthfulness and trust, so we define them too by reference to standards of justice and of courage. If A, a professor, gives B and C the grades that their papers deserve, but grades D because he is attracted by D's blue eyes or is repelled by D's dandruff, then he has defined his relationship to D differently from his relationship to the other members of the class, whether he wishes it or not. Justice requires that we treat others in respect of merit or desert according to uniform and impersonal standards; to depart from the standards of justice in some particular instance defines our relationship with the relevant person as in some way special or distinctive.

The case with courage is a little different. We hold courage to be a virtue because the care and concern for individuals, communities and causes which we feel requires the existence of such a virtue. If A says that he cares for some individual, community or cause, but is unwilling to risk harm or danger on his, her or its own behalf, he puts in question the genuineness of his care and concern. Courage, the capacity to risk harm or danger to oneself, has its role in human life because of this connection with care and concern. This is not to say that a man cannot genuinely care and also be a coward. It is in part to say that a man who genuinely cares and has not the capacity for risking harm or danger has to define himself, both to himself and to others, as a coward.

I take it then that truthfulness, justice and courage – and perhaps some others – are human goods in the light of which we have to characterize ourselves and others, whatever our private moral standpoint or our society's particular codes may be. For this recognition that we cannot escape the definition of our relationships in terms of such goods is perfectly compatible with the acknowledgment that different societies have different codes of truthfulness, justice and courage. Lutheran pietists brought up their children to believe that one ought to tell the truth to everybody at all times, whatever the circumstances or consequences, and Kant was one of their children. Traditional Bantu parents brought up their children not to tell the truth to unknown strangers, since they believed that this could render the family vulnerable to witchcraft. In our culture many of us have been brought up not to tell the truth to elderly great-aunts who invite us to admire their new hats. But each of these codes embodies an acknowledgement of the good of truthfulness. So it is also with varying codes of justice and of courage.

The inescapable character of such goods is a matter of the role in human life of the qualities that embody them, the virtues of being honest, just and courageous. A society that does not recognize these as virtues would necessarily lack the general features of human society. When we do find a culture, such as that of the Ik,[2] where this recognition is absent, we find something akin to a Hobbesian state of nature, not to a human society. This empirical confirmation of my thesis is strengthened by the discovery that the Ik once had genuine social bonds – and with them a recognition of the virtues – but lost these. To characterize this disaster by saying that the Ik lost *both* their social bonds *and* their recognition of the central virtues would be a mistake, for it would ignore the fact that to speak of the virtues just *is* to speak of certain key forms of social relationship. Virtues are parts of social structure, and moral philosophy and sociology ought not to be conceived of as distinct disciplines.

A certain tension exists between the moral history of our own culture as I understood it in the first section of this paper and the account of the virtues which I have just given. The first account emphasized cultural variability and fragmentation; the second has emphasized the irreducible moral content of culture. It may help to avoid the appearance of incompatibility if I immediately underline one point that I have already made and add to it another. The point already made is that the central invariant virtues are embodied in very different codes in different cultures and a shared recognition of these virtues is compatible with wide-ranging disputes about codes. The second and new point is that the central invariant virtues are never by themselves adequate to constitute a morality. To constitute a morality adequate to guide a human life we need a scheme of the virtues which depends in part on further beliefs, beliefs about the true nature of man and his true end. But about these matters cultures have of course varied and disagreed.

To understand how this is so, consider those qualities which have been considered virtues in some times and places and not in others: thrift, humility, charity, authenticity and friendship are cases in point. To understand these qualities as virtues requires an appeal to certain beliefs which have flourished in some human cultures and not in others. If we believe in the so-called Protestant work ethic – work, save, invest – (as practiced by Venetian Catholics and Spanish Jews as well as by German, English and American Protestants, pace Max Weber), the presence or absence of thrift will be a crucial fact about individuals. If we accept that certain forms of marriage and virginity are the will of God, then chastity will become an important quality. It is only against the background of such over-all interpretations of human existence that such qualities appear to be or not to be virtues.

In characteristic human cultures, therefore, the standard list of the virtues will include some items which derive their status from the part they play in all human life and other items which derive their status from some more particular set of beliefs or forms of understanding, which is restricted to some, perhaps to only one, form of culture and social order. What gives each moral vision its uniqueness and its integrity is not just a question of what items are included in its particular list of the virtues, but also of what criterion is invoked to justify the selection of just those items rather than others. So ancient Greek moral thought and practice is pervaded by the notion of human beings as having a distinctive nature, specified in terms of that 'work' (ἐργον) and that 'end' (τέλος) which distinguish human beings from gods, geese or granite. So the moral vision of the Renaissance man of action centers on the notion of 'virtu', a certain conception of human strength and energy. So the puritans of the sixteenth and seventeenth centuries gave an interpretation to the notion of a divinely ordained vocation which made both hard work and thrift virtues.

Such a criterion will determine not only which items are included in the list of the virtues, but also how they are ranked in importance. Courage is accounted a virtue in almost all societies, but it has a very different place in the heroic societies whose life is reflected in the Iliad or in the Icelandic sagas from its place in any Christian scheme.

The traditional medical virtues are clearly not to be derived in any simple way from the invariant human virtues. To count them as virtues we need to appeal to certain special beliefs about the specific kind of value we place on the preservation of human life, about the special character of the physician-patient relationship and about professional autonomy. The difficulty about the traditional medical virtues is two-fold: they have become problematic and they have become problematic in a culture which precisely lacks the means to solve moral problems. I have already indicated why I hold the latter thesis to be correct. Let me therefore turn to the former.

I want to argue that just as the traditional medical virtues had a special status, so those virtues have become problematic in a special way. There is a social process by which what have been virtues in one social and cultural context can become vices in another. I am not here referring to the process which I have already noted whereby what are *believed* to be virtues in one social order come to be *believed* to be vices in another, as the quality which Aristotle counted as a virtue under the name of magnanimity came to be considered by the early Christians the vice that is the counterpart to their virtue of humility. I am referring to a process whereby what actually were virtues turn into what actually are vices. Before I turn to the history of the traditional medical virtues, let me list a number of ways in which this may happen.

IV

The first is the case where a disposition valuable for its own sake, and valued for its own sake, as any genuine virtue must be (it may of course also be valued for further ends that it serves) comes to be valued only or primarily for its employment as part of a technique. Sometimes, for example, social workers are taught to become 'friends' with their clients in order to gain their confidence so as to manipulate them more effectively. Now it is of the essence of friendship as a virtue that one cannot become a friend from such a motive and with such an intention. What the social worker is being taught is to do and to speak as a friend would do and speak, but in such a way that what is produced is a counterfeit version and not authentic friendship. It is because the social worker's version is a counterfeit of the virtue and one calculated to deceive the innocent that it is a vice and not merely a neutral quality.

Aristotle when he discusses courage distinguishes true courage from military courage. Military courage is certainly not a vice, but because it is acquired and exercised as part of the technical training of a soldier it is not to be confused with a virtue which always belongs to a man *qua* man and not *qua* social worker, soldier or possessor of some other set of technical skills.

A second way in which a virtue may become a vice is when a change in the nature of the effects of a certain type of action transforms the character of that type of action. The giving of systematic poor relief to wage-laborers in England in the early seventeenth century, as a system institutionalized by a Poor Law and controlled by the local Board of Guardians, first had the effect of keeping wages low – for if the parish paid a near subsistence minimum the farmer could pay his laborers less than he would otherwise have had to; and then became a means to this end, so that the actions involved in poor relief were now informed by the intention of providing a stock of cheap wage labor. An unintended but very general effect led the action of assisting the poor to be transformed into the action of contriving their maximal exploitation. A virtue became a vice.

A third way in which a virtue may become a vice is when a quality valued for its own sake is made available for sale. Alexander Gerschenkron has described how in the initial stages of modernization and especially in the eighteenth century a new type of entrepreneur emerged whose transactions were based on much larger quantities of fixed and variable capital than previously and whose enterprises were essentially long-term.[3] Unlike his mercantile predecessors, he found himself having recurrent transactions with the same customers, debtors and creditors. Up to this point the rule, *Caveat emptor*, had tended to dominate commercial transactions; the entrepreneur did not need to convince those who were buying from him that he was reliable, if he would probably never trade with them again. The new entrepreneur, however, had to acquire a reputation for honesty and to acquire it he had to do by and large what honest men do. Would it therefore be true to say that entrepreneurs became honest? The answer, as Kant saw clearly, is 'No', and not only because 'honesty' became part of the technical skill of the entrepreneur. It is because what is exchanged for money and valued insofar as it can be exchanged for money cannot be a virtue. It follows, as the Marx of the 1846 manuscripts saw clearly, that free market capitalism has a striking tendency to destroy the virtues and a progressive tendency. It is not just that money is very bad for us; it is also the case, as Robin Farquarson put it, that "Marijuana is not addictive; money is."

Fourthly, a virtue may become a vice or simply a non-moral quality by a change in its relationship to a role which it partially defines. In traditional, patriarchal societies reciprocal loyalty between superiors and inferiors is central to the social bonds. To value loyalty as a virtue just is to value those bonds. But when the virtue of loyalty is invoked in modern corporate organizations it is characteristically part of an attempt to undermine the impersonal, bureaucratic standards of such organizations and to substitute adherence to persons for adherence to rules. Such adherence may be a good or bad thing depending on the circumstances and the nature of the relevant persons and rules; but it is certainly not a virtue as such.

We have then at least four types of case in which virtues can become vices, or if not vices, at least problematic qualities: the type of case where what was a virtue is transformed into part of a set of technical skills, that where a change in the effects of the practice of a virtue transforms in time the intentions with which it is practiced, that where what was a virtue becomes a marketable commodity, and that where a change in the structure of roles changes the character of the human qualities involved in the role playing.

V

I now want to consider if and how far the traditional medical virtues have turned to vices. I want to begin by considering three social presuppositions of the practice of the traditional medical virtues. The first is technological. The practice of medicine has for most of its history been carried on in societies where human life is immensely fragile and vulnerable and where the technical means to safeguard it have been very limited. High infantile mortality rates, low expectations of life for surviving adults, extremely limited predictive powers in framing prognoses, all underlie the ordering of medical priorities embodied in different versions of the Hippocratic Oath. Medicine would have been a quite different form of social practice if *either* life was to be preserved only if health could be restored *or* life was to be preserved only if grave pain and suffering were to be avoided *or* health was to be restored only if in so doing pain and suffering were not to be increased. That ordering of medical priorities which places a supreme value on life is made more intelligible by considering the social background which it originally presupposed.

A second presupposition of the practice of the traditional medical virtues was the existence of a shared and socially established morality. The physician could assume that the patients' attitudes towards life and death would be roughly the same as his own, and vice versa. Hence the patient in putting him or herself into the hands of his or her physician could feel that he or she was not relinquishing his or her moral autonomy.

A third presupposition of the practice of the traditional medical virtues was that the activities of the physician or surgeon took place within a given social order, but were not themselves able to shape or be responsible for shaping that order. Medicine could not be understood in its traditional perspective as a social practice competing with other social practices for scarce resources and offering debatable criteria for their distribution.

None of these presuppositions is now warranted and it is social change that has destroyed their warranty. Technological change has made of the preservation of human life a very different issue. Moral change has made of the trust which the patient ought to express in the physician a very different issue. Changes in the scale and the cost of medical care as well as political and economic change in society at large have made the distribution of medical care into a very different issue. In each case what was a virtue has become at best problematic, at worst a vice. Consider once more the ways in which virtues become vices.

There is first the case where the effects of a practice change so that the character of the relevant actions change. This is what happened to the medical practice of making the preservation of human life an overriding goal. Consider two kinds of change. It is now the case, as it used not to be, that this goal involves the systematic preservation of the old long after they can function as genuine human beings. It is now the case, as it used not to be, that this goal involves systematically increasing the proportion of hopelessly crippled infants and helplessly decaying old people to healthy adults and children. Any agent who knowingly participates in producing such effects systematically, as many physicians do, does great harm and wrong. What was a virtue has become a vice, but not an unproblematic vice. For the physician now finds himself in a tragic dilemma. Consider the case of recently born crippled infants where heroic efforts may preserve *either* a needless bundle of distorted and suffering nerves and tissues *or* – sometimes against all probable calculation – a human child, physically imperfect but with real potential, perhaps even a Helen Keller. (I consider the case of infants rather than of the old, because the collapse of the extended family has left most of us with a deep inability even to approach the problems of the old, an inability institutionalized in the way we, as a society as well as individuals, treat them.) Any rule which relieves the physician of the burden of extending suffering uselessly imposes on him the burden of taking innocent life wantonly; and no rule would be worst of all.

What has happened to place physicians in this dilemma is the result of the coincidence of two distinct histories of moral change. In the society at large our fragmented inheritance has resulted in abandoning us to a secular, liberal pluralism which leaves us resourceless in the face of moral problems; in the history of medical practice a change in its presuppositions has rendered what was virtuous vicious and what was unproblematic problematic. Thus parts of medical practice became morally problematic precisely at a time when we have minimal resources for the solution of moral problems.

As with the first of the three traditional medical values, so also with the other two. The trust which defines the relationship of patient to physician was based upon the presupposition of a shared, established morality. The physician could have a reasonable assurance that his patients' beliefs about suffering, death and human dignity were much the same as his own; the patient could have a reasonable assurance that his beliefs would be respected. But in a liberal, pluralist moral culture the patient knows, not only that the traditional basis for this assurance is now missing, but that the physician has no special resources for the solution of the moral problems which arise in the course of a relationship to a patient. The parent of a helplessly ill child or a helplessly old person cannot know that the physician wills their good, because they cannot know what his conception of good is. Once again the physician is in a tragic dilemma: the invitation to trust which was once a sign of virtue becomes a sign of something else. The change in the structure of role-playing has changed the quality of the actions. A virtue has in a characteristic way become a vice. But the physician has no easy way out. The whole nature of medical care is almost unimaginable without a context of mutual trust; to simply abandon that mutual trust, because it is no longer warranted, would be destructive. To try to maintain it in its traditional forms is equally dangerous.

It is of course in this situation that market relations become significantly obtrusive in medical practice. Differential treatment is offered for differential reward; access to medical care is radically unequal. Here again the physician is, like everyone else, in a situation which he cannot escape. The demands of social justice and the demands of the physician for autonomy are in radical conflict. If members of the medical profession choose certain forms of specialization in research or in practice, they thereby determine the availability of certain patterns of medical care. If the freedom of physicians is safeguarded, the equal rights of citizens will be flouted. So the autonomy of the medical profession becomes a social vice, while the freedom of the physician remains an important value. Once again we have a dilemma which is almost intolerable.

VI

Hegel spoke of tragedy as "the conflict of right with right;" what makes any protagonist's situation tragic is that he inevitably has to choose between wrong and wrong. It is with this in mind that I have spoken of the physician's moral dilemmas as tragic. The moral resources of his culture, of our own culture, offer no solution for him. What matters most in a period in which human life is tragic is to have the strength to resist false solutions. The characteristic temptation of the modern world is utilitarianism. For utilitarianism in all its versions aspires to provide a criterion, a way of judging between rival and conflicting goods to maximize utility. But the goods and the rights which define our contemporary conflicts are incommensurable. There is no higher criterion. There is no neutral concept of utility.

The medical profession ought not therefore to look for solutions to philosophical theorizing; what philosophy has to tell them is precisely why they cannot hope for solutions. For a philosopher to try to go beyond this would be for him to misunderstand either the present situation or the scope and limits of his discipline. A philosopher offering positive moral advice in this situation would be a comic character introduced into a tragedy. Imagine Socrates introducing himself with advice for Antigone or Creon, or Plato trying to counsel Philoctetes, Neoptolemus and Odysseus. Yet to understand even this is perhaps to transform the perspective in which the moral problems of medicine are viewed; and such a transformation can only be effected by philosophy.

Boston University,
Boston, Massachusetts

NOTES

[1] Charles L. Stevenson, *Ethics and Language* (New Haven: Yale University Press, 1944).
[2] Colin M. Turnbull, *The Mountain People* (New York: Simon and Schuster, 1972).
[3] Alexander Gerschenkron, "The Modernization of Entrepreneurship," in *Modernization: The Dynamics of Growth*, ed. by Myron Wiener (New York: Basic Books, 1966), pp. 246 f.

SAMUEL GOROVITZ

MORAL PHILOSOPHY AND MEDICAL PERPLEXITY: COMMENTS ON "HOW VIRTUES BECOME VICES"

Professor MacIntyre has given us an eloquent, insightful, and persuasive presentation. In addition, unlike many other eloquent, insightful and persuasive presentations, it is important, addressing as it does not only fundamental issues in moral philosophy and in the analysis of medical practice, but also the question of whether there is any significant gain that can be reasonably expected as a result of bringing philosophy to bear on the morally troubling problems that arise in medical practice. We should all be grateful for his provocative analysis. I agree with much – perhaps with most of what is basic – in Professor MacIntyre's remarks about medicine. But I believe that his conception of what it is reasonable to expect of moral philosophy is wrong in important ways, so that in consequence his final thesis about the relationship between philosophy and medicine is wholly unwarranted. And further, I will argue, unlike many other theses that are wholly unwarranted, it is pernicious, having as its consequence the view that philosophers ought not attempt to do what I claim it is both possible and crucial for them to do. The essence of my response will be quite simple: I grant that philosophy cannot provide what MacIntyre seems to wish of it, but I hold that it can nonetheless provide important goods that MacIntyre fails to acknowledge. I will address exclusively, and, I fear, with greater brevity than his essay merits, what I take to be MacIntyre's main conclusions about moral philosophy and medical perplexity – although there is much more that, given more time, I would like to address.

We should have no confusions about what those main conclusions are: MacIntyre articulates them quite clearly. Referring to the morally troubling questions in medicine, he says, "We have no rational method available for reaching a conclusion in these questions." Rejecting the possibility that this unhappy state of affairs is due either to the special character of these problems or to the general character of moral argument, he places the blame squarely on "the peculiar character of moral debate in our liberal, secular, pluralist culture." And in the end, he con-

H. T. Engelhardt, Jr. and S. F. Spicker (eds.), Evaluation and Explanation in the Biomedical Sciences, 113–121.

cludes, "The medical profession ought not... look for solutions to philosophical theorizing; what philosophy has to tell them is precisely why they cannot hope for solutions."

I reject this characterization of what philosophy has to tell the medical profession. But I believe I understand how Professor MacIntyre comes to hold it. If we accept the view that a criterion of adequacy for moral theory is that it enables us to resolve any moral dilemma, and if we then observe that there seems to be no moral theory that performs such a function, we can validly conclude that there is no adequate moral theory. It then follows that there is no moral theory adequate to resolve the physician's dilemmas in particular, and further that physicians ought not to look to moral theory for such solutions. If the absence of an adequate moral theory is due to our liberal, secular, pluralist culture, rather than to some basic feature of moral argument, so be it; it is an absence all the same.

There is some evidence that Professor MacIntyre holds such a view of moral philosophy. He speaks of "a morality adequate to guide a human life," argues that virtues and vices can be distinguished only against the background of an "over-all interpretation of human existence," and claims that our "fragmented" moral inheritance has left us "resourceless in the face of moral problems." I have a sense here of an attitude toward moral philosophy that reflects a quest for moral certainty, much as Descartes sought epistemological certainty. MacIntyre would have us believe that because we have no coherent, all-encompassing moral perspective, we are morally resourceless. But the quest for moral certainty makes no more sense than the Cartesian quest. Rationality demands that we accept less than perfect evidence as being good enough to warrant belief – because there is no perfect evidence. Similarly, there is no perfect assurance of moral rectitude, and hence moral courage demands that we accept less than perfect practical reasoning as being good enough to warrant action. In ethics as in epistemology, uncertainty can have its place, without undermining the enterprise in its entirety.

Let us look at how MacIntyre supports his claim of moral resourcelessness. We are given three pairs of arguments, each with plausible premises, valid inferences, and incompatible conclusions. The first pits our moral intuitions about universality against our respect for bodily integrity. The second opposes health and perhaps even life to honesty. The third reveals social welfare again in conflict with individual freedom. I agree that such dilemmas hold the makings of tragedy.

There is no rational method for resolving these disputes, MacIntyre argues, because those who accept the premises of rival arguments share no common moral ground. And this, presumably, because we have "no rationally defensible concept of man's true end or of an essential human nature." Instead, all we have in our moral arsenal is a "collection of fragments" – moral principles available for independent assertion from a variety of moral points of view.

I believe that this characterization of our moral plight, on which characterization the development of Professor MacIntyre's argument essentially depends, fails to do justice to what we know about moral reasoning. In order properly to challenge his conclusion about what philosophy can say about the moral problems in medicine, I must therefore provide a different perspective on that plight.

The moral principles that we espouse are indeed in conflict. We are all well versed in how one must sometimes lie to keep a promise or inflict pain to prolong life. In some cases, no moral dilemma is involved. We pause to save the drowning child without distress at the fact that we will thereby be somewhat later for dinner than we promised. Yet we retain our belief that it is good to keep promises. But sometimes the conflict does involve dilemma, and the cases drawn from medicine are paradigmatic of this sort. In such cases not only do principles of value conflict, they conflict in such a way as to cause substantial distress. One does not know which of the conflicting principles to honor; instead one suffers the anxiety of moral uncertainty. Can we reasonably blame the conflicting principles themselves, or our lack of a global moral view in terms of which to resolve the conflicts? Is it true that we are resourceless; that the proponents of rival moral arguments share no common moral ground? I think not.

The moral principles we honor are not simply a random collection of ancient ethical artifacts. On the contrary, they are articulations of values that we accept – which is to say, descriptions of goods that we want. Among the goods that we want are honesty, respect for bodily integrity, individual freedom, health, and many more. The principles corresponding to these goods survive because these are values we do hold, and the importance we attach to each principle reflects the centrality of the corresponding value. These principles, it is important to recall, pertain to actions; moral judgment, after all, is primarily judgment about how people ought to act. But man as a moral agent faces a morally recalcitrant world. I do not refer to the behavior of others. I refer, rather, to the fact that in many ways that matter a great deal, we simply cannot have what we want. There is nothing contradictory about the concept of a world in which the relief of suffering and the prolongation of life are never in conflict. There is nothing in logic to preclude a world in which patients are never at risk of being injured by the truth. Indeed, there is no conceptual flaw in a world of perfect health. But these are not our world.

Man as a moral agent is all too often like the hungry dieter confronted with a hot fudge sundae. He can imagine a world in which his hunger would be satisfied with a single bite; he can contemplate a world in which such delicacies are not fattening. But in this world, his desires are in conflict. Ambivalence is his state, dilemma his burden, regret his likely companion. Yet there is nothing inappropriate in *wanting* to be fed but not fat. He is thwarted by the empirical world, not by any incoherence in his desires.

What, then, of the physician who respects life and respects a pregnant woman's bodily integrity, who is convinced by argument (1a) that he ought not to perform a requested abortion and is convinced by argument (1b) that he ought to perform it? Must he flit like Stoppard's jumpers from stance to stance, unable to find any basis for landing on either side of the issue? Like our hungry dieter, he will be ambivalent, and will recognize in the conflict he faces the moral recalcitrance of the empirical world. But he is not necessarily resourceless, any more than the hungry dieter is resourceless. For the dieter can acknowledge that delicious food is good and that fattening food is bad, can recognize that the object of his judgment bears many attributes including those of being delicious and being fattening, and can then adopt a broader perspective in his quest for decision. He can reflect, for example, that the aspirations that inform what he presently sees as his rational life plan would be advanced more substantially by weight reduction than by immediate gastronomic gratification, and thus may abstain – with regret at what he thereby misses conjoined with satisfaction at what he thereby gains. Or he may decide that his deprivation has been so sustained and his progress so substantial that the incorporation of inspiring ingestible incentives from time to time is both deserved and prudent. In either case, he gets off dead center by relating the two conflicting arguments to a broader prudential context in terms of which he can override one in favor of the other. He takes into account both of the conflicting arguments, and reconciles the conflict by judging them against an overriding purpose – not, to be sure, an overriding purpose that informs all human actions; merely a purpose that is overriding with respect to the two alternative actions under consideration. But that is overriding enough.

Can the troubled physician not do the same, resolving his dilemma in terms of a broader moral perspective that enables him to choose between two troubling alternatives? That he can is precisely what MacIntyre denies. Recall his explicit assertion that "those who accept the premises of rival arguments share no common moral ground." Yet the rivals in question may well be – indeed, often are – just two different facets of the moral reasoning of one individual faced with a choice. Why is there no common moral ground? Presumably because the conflicting principles are unrelated fragments from the past, cut adrift from the cohesive foundation that can be provided only by a sense of man's true end or essential nature.

I believe a different description of our moral circumstances is needed here. The agent faced with moral dilemma who is thus in internal moral conflict, and the separate moral adversaries too, may find no common ground in the premises of the moral arguments they accept. But MacIntyre has offered us no argument that shows they cannot find common moral ground by expanding the scope of their deliberations. Consider a case in point. A child in the advanced stages of Hurler's disease falls ill with a life-threatening infection. His disease is a grotesque, degenerative, terminal illness. His infection is readily treatable. Should the physician cure the infection, thus saving the child's life for a brief period of continued degradation, frustration, and suffering? Opinion *A* argues for treatment, citing the invariant value and sanctity of life and the physician's consequent obligation always to strive to preserve it. Opinion *B* argues against treatment, citing the value of kindness, of welcoming the benevolent intervention of natural events in a chilling human tragedy. Have they truly no common moral ground?

Let us press *B*, and ask why he opposes medical intervention to thwart the infection. It is plausible to assume that he will cite such factors as the

pain suffered by the patient and his family, the utter hopelessness of
Hurler's disease, the draining costs, psychic as well as monetary, of keeping the patient alive, and all the rest. Under further questioning, he may
allow that these outcomes are undesirable because they constitute or
cause agonizing experiences for all concerned, including the patient. Allowing the patient's life to end, on the other hand, will minimize such
experience. Now let us press *A*, and ask him why he holds life to be always
worth preserving. There are many answers he might give. He might argue
that where there is life, there is hope, suggesting that a cure for the disease
may be announced at any moment. He might argue that medical research
always stands to gain from considering the seriously ill. The longer the
patient survives, the more data are available. He may assert that life
simply is of intrinsic and overriding worth, and thus should be preserved
no matter what. Each of these three responses, of course, reflects a significantly different moral basis for the view that life has value. Finally,
he may observe that life has value because it is a prerequisite for all else
that is of value – in particular, it is a necessary condition for joy, pride
in accomplishment, service to others, a sense of beauty, indeed, even for
self-indulgence or self-pity. In short, it is a precondition for experience,
without which there is no intelligible sense in which value exists at all.

Now I readily admit that the conversation might not go like that. But
assume for the moment that it does. Then if *A* can be convinced that there
is no significant prospect of the patient having valuable future experiences
or contributing to valuable future experiences of others, whatever such
experiences are taken to be, then *A* may relinquish the view that *this*
patient's life is valuable at this time, holding now to the view that life
for the most part, precisely because it is the precondition of the experiences that constitute the basis of value, is of overriding importance. *A*
and *B* may then agree.

Again, I do not argue that the story *must* follow this course – only that
it might. And if it does, then we have an instance wherein moral conflict
is resolved precisely because the adversaries have come to see that there
is, or have come to adopt, a common moral ground after all, in spite of
their having begun with opposing views derived from ostensibly irre-
concilable moral positions.

It is important to be clear about how this resolution has proceeded.
The conflict has not been dispelled in any *ad hoc* way. Rather, three

systematic steps have led to a mutually acknowledged general principle. First, the disputants examined the reasons and presuppositions that supported their rival positions. Second, they identified in those presuppositions the common moral ground that experiences alone have intrinsic value. Third, they abstracted from that common moral ground to a principle that resolves the present case: life is valuable only because of, and therefore only in the presence of, valuable experience or the possibility of valuable experience. This principle is generalizable, however, to other cases, and constitutes a more basic level of moral judgment for both disputants than their original rival premises.

Of course, nothing in this example shows the principle to be incontrovertible. It is not self-certifying, nor has it been shown to be the conclusion of a valid argument with incontrovertible premises. Thus, even if we are inclined to accept the principle, we can maintain a residual uncertainty about whether action based on it is right *sub specie aeternitatis*. If MacIntyre's point is simply that such certification is unavailable, I quite agree. But that, of course, has nothing whatever to do with the ability of philosophy to find common moral ground between disputants.

What I am claiming here is that a philosophical inquiry into the reasons that underlie conflicting moral judgments about a disputed case can facilitate the resolution of the conflict by placing those reasons in the broader context of shared values. I do not claim it can always be done, but I do dispute MacIntyre's claim that it can never be done – that we are literally resourceless. One reason why it might be impossible in a given case is that a common moral ground may indeed be absent and unachievable. Still, if philosophy can help to find the common ground where it does or can exist – as is surely true in many difficult cases – then there is much that the medical profession can reasonably expect from moral philosophy. Further, it would be a mistake to assume that because there seems to be widespread disagreement about moral issues in our culture, such bases for agreement are extremely rare. MacIntyre himself has argued well that ostensible cultural diversity concerning values lends little support to ethical relativism.

Of course, on my account there will be unresolved conflict and enduring moral uncertainty. I simply do not ask that moral philosophy eliminate moral dilemma before I will acknowledge it as useful and important. Rather, I believe that moral philosophy arises precisely out of moral dilemma, and persists primarily because moral conflict endures. I am willing to respect it for what it can do instead of lamenting what it cannot.

Virtues *can* become vices; the prolongation of life is indeed a case in point. That traditional medical value has become more problematic both because of our increased ability to prolong life in the face of an unhappy prognosis and because of our increased understanding of why it is that life is valuable. But this does not happen, as MacIntyre claims, "in a culture which precisely lacks the means to solve moral problems." The physician does face a tragic dilemma, for example, in dealing with the tenuous lives of certain patients with severe disabilities. But I have some observations to add about the ingredients of tragic dilemmas. A child may be torn between Sesame Street and The Electric Company, yet there is no tragedy involved because the issue is simply not sufficiently important. A senior may be torn between Stanford and Johns Hopkins in choosing a medical school, but again there is no tragedy. Although the issue is of major importance, there is no sense in which there exists a correct choice that the student tragically may fail to discover. Thus, there is no sense in which there is a profound seriousness to judging wrongly. What makes the physician's dilemma tragic is that the importance of the issue is so often conjoined with a sense of the profound seriousness of making a mistake. Yet the sense of tragedy can be overdone. Virtue does not lie in saintliness any more than knowledge lies in certainty. Physicians can act with moral integrity even in the face of moral uncertainty – although doing so adds new demands to those of medical integrity.

Philosophy can help articulate those demands and help point the way toward meeting them. It can help us to understand the relations between our actions and our aspirations, and can teach us to accept as necessary the moral conflict that inevitably arises from the clash between what we want and what the empirical world will yield to us. As a kind of cognitive therapy, it can help us to identify and understand what we really do value. Moreover, there is the possibility that, in addition to the benefits of philosophical analysis, philosophy can provide a positive moral theory that is based on the moral content of a liberal, secular, pluralist culture. In sum, it can tell us not, as Professor MacIntyre claims, why we cannot hope for solutions, but rather what sorts of solutions we can reasonably hope for. And it can help to guide us toward them. Of course, moral philosophy alone can not provide solutions to moral problems any more than law alone can eliminate legal problems or economics alone can eliminate economic problems. We must always reach beyond philosophy in addressing problems in the world. But we should be wary of reaching without it.

Professor MacIntyre tells us that "what matters most in a period in which human life is tragic is to have the strength to resist false solutions." No period of human history has been devoid of tragedy; none, I'm sure, ever will be. Human life is always tragic, and false solutions should always be resisted. So, too, should false despair.

University of Maryland,
College Park, Maryland

SECTION IV

CONCEPTS IN MEDICAL THEORY

H. TRISTRAM ENGELHARDT, JR.

THE CONCEPTS OF HEALTH AND DISEASE

Health and disease are cardinal concepts of the biomedical sciences and technologies. Though the models of health and disease may vary, these concepts play a defining role, indicating what should and what should not be the objects of medical concern. The concepts are ambiguous, operating both as explanatory and evaluatory notions. They describe states of affairs, factual conditions, while at the same time judging them to be good or bad. Health and disease are normative as well as descriptive. This dual role is core to their ambiguity and is the focus of this paper. In this paper I shall examine first the concept of health; second, the concept of disease; and third, I will draw some general conclusions concerning the interplay of evaluation and explanation in the concepts of health and disease.

I. HEALTH

Health is a normative concept but not in the sense of a moral virtue. Though health is a good, and though it may be morally praiseworthy to try to be healthy and to advance the health of others, still, all things being equal, it is a misfortune, not a misdeed, to lack health. Health is more an aesthetic than an ethical term; it is more beauty than virtue. Thus, one does not condemn someone for no longer being healthy, though one may sympathize with him over the loss of a good. Further, it is not clear exactly what is lost when one loses health.

The norms of health are difficult to compass within one homogeneous concept, in particular within an independent definition which does not define health negatively, as the absence of disease. The World Health Organization attempted a positive definition, that "health is a state of complete physical, mental and social well-being and not merely the absence of disease or infirmity."[1] But such a definition of health packs the ambiguity of the concept of health into the ambiguity of a concept of well-being. Further, this concept of well-being suggests the notion of a satisfactory lifestyle, including successful adaptation to one's environ-

H. T. Engelhardt, Jr. and S. F. Spicker (eds.), Evaluation and Explanation in the Biomedical Sciences, 125–141.
All Rights Reserved. Copyright © 1975 by D. Reidel Publishing Company, Dordrecht-Holland.

ment. Yet, even here the norms are obscure. What is a good adaptation? Is a good adaptation possible in a complex industrial society for those with I.Q.'s of less than 80? Are such persons ill? Further, if health is a state of complete physical, mental and social well-being, can anyone ever be healthy? Does health become a regulative ideal, one to which one strives, but which one can never fully achieve? On the other hand, if no one is truly healthy, is everyone ill? Are health and disease exclusive or overlapping concepts?

These quandaries arise primarily out of the evaluatory, not the explanatory, dimension of the concepts of health and disease. Health could, for example, be defined as the ability to perform those functions which allow the organism to maintain itself, all other things being equal, in the range of activity open to most other members of its species (e.g., within two standard deviations from the norm), and which are conducive toward the maintenance of its species. This, though, is to forego mention of why one might be interested in certain types of well-being (except for the commitment to the survival of the species). Also, it is not clear whether, within such a concept, degenerative processes generally distributed in the population, which appear after the reproductive years, could count as diseases. Finally, if health is to encompass issues raised by particular diseases, the concept may lose its unity. The attempt to understand the concept of health thus brings us to the concept of disease, suggesting that the concept of health may have as many nuances as there are diseases, and that it may be derivative from these particular disease concepts.

II. THE CONCEPT OF DISEASE

The concept of disease is used in accounting for physiological and psychological (or behavioral) disorders, offering generalizations concerning patterns of phenomena which we find disturbing and unpleasant. The concept of disease is a general scheme for explaining, predicting, and controlling dimensions of the human condition. It grades into other concepts which are political, social, educational, and moral. The difference between the concept of disease and these other concepts, and the similarity of the various models of disease is complex and problematic. It is not even clear that all the models of disease fall within a single genus. Perhaps the concept of disease indicates a family of conceptually consanguineous notions. That is, the concept of disease may be a basically heterogeneous concept standing for a set of phenomena collected together out of diverse social interests, not on the basis of the recognition of a natural type or a common conceptual structure. Disease would then be whatever physicians in a particular society treat, rendering circular the definitions of disease and medicine.

It is worthwhile distinguishing, somewhat stipulatively, disease and illness. One can be ill, feel bad, feel under the weather without explaining such phenomena in terms of disease models. The concept of disease competes with other concepts, from demonic possession to simple exhaustion. And, on the other hand, one can have a disease without being ill, as in the case of Alvan Feinstein's concept of lanthanic diseases.[2] A person can, for example, have carcinoma of the lung diagnosed before he is ill. To speak of diseases is to make an explanatory move and at the very least to convert an illness into a syndrome, to recognize a cluster of phenomena as a disease pattern.

The concept of disease acts not only to describe and explain, but also to enjoin to action. It indicates a state of affairs as undesirable and to be overcome. It is a normative concept; it says what ought not to be. As such, the concept incorporates criteria of evaluation, designating certain states of affairs as desirable and others as not so. It delineates and establishes social roles such as being sick or being a physician, and it interconnects these roles with a network of expectations structured by rights and duties.[3] The concept is both aesthetic and ethical, suggesting what is beautiful and what is good. By terming something diseased, one indicates that the state of affairs is both naturally ugly as well as one that imposes some obligations and relieves others.

The concept of disease is thus freighted with important ambiguities. These ambiguities are, as well, bound to what has been termed the physiological and ontological concepts of disease, to the levels of abstraction involved in disease models, and to the nature of particular models of disease, such as the medical and psychological. I wish to introduce the physiological and ontological concepts of disease from a primarily typological, not historical, point of view. They represent two general ways of talking about disease. Historically, they developed out of disputes whether disease was the result of the malmixing of humors, or was due to the entrance of a disease entity. The dispute roughly was whether diseases were primarily relational and contextual in character, or in some sense substantial things. These different ways of talking about disease are still apparent.

There is an important ambiguity in the significance of ontological concepts of disease. The *ens*, the being of the disease, can be variously understood as either a thing, or a logical type, or both. Medical ontology in the strong sense refers to views in which disease is conceived of as a thing, a parasite,[4] in contrast with "Platonic" views of disease entities in which diseases are understood as unchanging conceptual structures. In the strong sense a "disease entity" is a disease thing, a material, invading agent of disease.[5] This strong sense of ontology involves a commitment to hypostatization, an attempt to reify disease. For example, Paracelsus, whom Pagel describes as the prototypical ontologist, opposed the humoral pathologists who held that "the sick individual determines the cause and nature of disease." In contrast, Paracelsus taught "the 'ontological' view in which diseases are regarded as entities in themselves distinguishable by specific changes and causes." In this view, it was "the individual disease that conditions the patient and manifests itself in a characteristic picture."[6] This analysis suggested that specific therapies should be sought for specific diseases. Or, more fundamentally, it advanced the notion that diseases had specific characters, and would thus respond to specific therapies. Moreover, his view suggested a distinction between "symptomatic" and "aetiological" therapy, therapy aimed at the results and therapy aimed at the cause of disease. It was this ontological, aetiological concept of disease which indicated the possibility of classifying therapies according to their focus on the specific causes of diseases. Further, specific diseases led to specific local organ changes being sought by van Helmont and others, and thus to the beginning of modern pathology.[7] The identification of disease with particular causes, therapies, and, finally, with localized pathological changes in organs is the focus of this viewpoint.

The ontological concept of disease spans the meaning of disease from Paracelsus' concept of the disease as a parasite, to concepts of contagion as found in Harvey,[8] to modern bacteriological concepts in which illness is variously identified with infectious agents. Bacteriology finally vindicated Sydenham's faith in the possibility of specific remedies for specific diseases. As Knud Faber put it, diseases "came to be viewed from an

etiological point of view, and the efforts of clinicians were directed towards... a nosography founded on the morbific causes."[9] The picture
emerged of the host, the environment, and the disease agent as the elements of disease with the accent falling heavily on the agent of disease,
particularly as an infectious agent. As Rudolf Virchow remarked, this
"idea of particular, parasitic disease entities is without doubt ontological
in an outspoken manner."[10] This ontological view involves as well a
confusion of the cause of illness and the disease itself, as when the disease,
tuberculosis, is identified with Mycobacterium tuberculosis. Virchow saw
this as beginning with the discovery of microorganisms and presupposing
a "hopeless, never-ending confusion, in which the ideas of being (*ens
morbi*) and causation (*causa morbi*) have been arbitrarily thrown together."[11] In this, Virchow was undoubtedly correct. He, though, was
himself also, by his own admission, "a thoroughgoing ontologist," who
identified the pathological findings as the *ens morbi*, the disease entity.[12]
Though he eschewed one form of reifying nosology, he embraced another. Disease became specific pathological changes in specific cells.

But the move to reification is only one dimension of the ontological
thesis concerning disease. The other dimension is bound to a judgment
concerning the nature of the disease pattern, the constellation or cluster
of the signs and symptoms which form the character of a disease. In
ontological theories, these characteristic disease patterns are interpreted
as enduring disease types often without an immediate connection to a
particular theory of material disease entities. It is an ontological question
(in the philosophical sense) of the reality of disease types, one that asserts
that they have a being beyond their particular instantiations. This interpretation of illnesses as portraying natural or essential disease types suggests an almost Platonic construal. In this view, courses of illnesses more
or less fully achieve a natural type. Classical cases of a disease are thus
perfect instantiations of a disease type and atypical cases imperfect real-
izations of a disease reality which exists as a natural, logical possibility.
"Ontological" in this sense refers to claims of a reality for diseases existing
apart from their embodiment in actual illnesses, which illnesses may be
"atypical."

It is not clear that anyone held a fully realist view of diseases, but it is
a viewpoint presupposed in the ordinary discussions of hospital wards
where reference is made to "cases of X disease" in a way parallel with

phrases such as "instances of X idea." To talk of illnesses achieving certain constant and real patterns, best illustrated by classical cases, and obscured by atypical cases, is to talk in a realist mode. To a point, it is illustrated by such classical nosologies as that of Francois Boissier de Sauvages (1706–1767), *Nosologia Methodica*.[13] Even with his empirical dedication, symptom constellations were for Sauvages types of diseases. Pinel's *Nosographie philosophique*[14] distinguished the varying symptoms of diseases from the "essential fevers," and attracted Broussais' classical critique of ontological theories of diseases. "One has filled the nosographical framework with groups of most arbitrarily formed symptoms... which do not represent the affections of different organs, that is, the real diseases. These groups of symptoms are derived from entities or abstract beings, which are most completely artificial ουτοι; these entities are false, and the resulting treatise is ontological."[15] As Peter Niebyl indicated, Broussais' objection was against abstraction, not against the specificity of disease entities.[16] The moral of the story is that there are strong Platonic, realist tendencies in talk about diseases, and that the tendencies have met opposition from anti-realist quarters bearing a resemblance, if only distant, to the old humoral theories. When viewed in contrast to such anti-realist theories, ontological theories of disease indicate more "the ever-recurring craving of... clinicians for... fixed categories of diseases,"[17] than an attempt to reify disease concepts, which I have called medical ontology in the strong sense. The ontological sense of disease thus spans a range of significance from the concept of a specific logical entity to a specific material entity.

The traditional viewpoint contrasting with that of the ontologist has been the physiological or functional viewpoint. Lord Cohen drew the contrast as between a Platonic, realist, rationalist versus a Hippocratic, nominalistic, empirical view of disease.[18] The argument against ontological concepts was, as Wunderlich realized, an argument against the logical blunder of confusing abstract concepts with things, "presupposing them as actually existing and at once considering and treating them as entities," as well as against "models of diseases which contain no truly essential feature... [and] to which we only by way of exception or by using compulsion, find an example in nature."[19] But it was not a denial that there are patterns of disease processes. Essential disease types are not the same as basic laws of physiology or pathophysiology.

Those arguing for a physiological concept of disease had at least three points to make, which individual nosologists made more or less completely. First, they wished to secure the concept of disease as a general, not a specific notion. That is, they wished to make diseases functions of the general laws of physiology rather than functions of the more particular laws of the pathology of specific diseases. Yet, second, they wished to argue for a greater appreciation of the individuality of illnesses so that every particular disease-state could be understood in terms of its particular departures from general physiological norms. Heinrich Romberg put it this way, "And we do not regard the mere placing of the disease under this or that rubric as the final aim of diagnosis.... The most important thing remains the determination of the degree that the individual human is injured by his malady, and which cause has produced the momentary disorder." [20] Third, they wished to avoid the metaphysical and logical muddles of ontological concepts of disease. Diseases were not things nor were they perduring types of pathology. Rather, for the physiological or functional nosologists, diseases were more contextual than substantial, more the resultant of individual constitutions, the laws of physiology and the peculiarities of environment, than the result of disease entities. The physiological nosologists were closer to the Hippocratic appreciation of the nexus of airs, waters, and places.

The dispute between ontological and physiological theories of disease turns centrally on the ontological and logical status of disease entities. Ontological theorists framed views within which diseases could be appreciated as specific entities. Physiological theorists framed views within which diseases could be appreciated as particular deviations from general regularities. In the first case, the accent of reality fell upon the disease; in the second case the accent fell upon the individual and his circumstances, including the laws of physiology. There is a temptation to see this contrast as between a realist and a nominalist construal of the meaning of disease. There are surely strongly nominalist leitmotifs in physiological and functional theories of disease. There is an emphasis upon the individual, not the disease, as the reality in the illness. But though there is a sympathy with nominalism, there is no commitment; physiological theorists have been willing to speak of diseases and accord them a conceptual reality. For example, Ottomar Rosenbach's term, "ventricular insufficiency," indicated a disease state which has a real universality, though no reality apart from the instances of which it is the common property. It is a similarity between processes but not a thing, nor a disease entity with a unique cause. It is a resemblance common to a family of processes, not just a family of processes collected together out of various resemblances.

In the end it is because of the need for universality that both ontological and physiological disease theorists required more than a nominalism, though less than a full-blown realism. Progress from syndromes to disease entities helped secure the possibility of diagnosis, prognosis, and therapy, or, more generally, medicine's explanation, prediction, and control of reality. More than the mere ability to name similar objects, one needed as well to indicate a common structure in reality, even if that structure was only the physiological norms from which diseases were departures. Mere syndromes are, as the name implies, the running together of symptoms and signs. They are a constellation of phenomena without a nomological structure to bind the signs and symptoms in a fashion to provide a model for explanation. Treatment and prognosis concerning mere syndromes are empiric in the derogatory sense of a maneuver based on correlations but devoid of an account of the relation between the phenomena correlated. Disease entities offered a level of abstraction that could bind together the signs and symptoms in an etiological matrix. In particular, reifying medical ontological theories could treat diseases as the substances which bear the signs and symptoms, the accidents of the underlying essential reality. Thus identification of phthisis with Mycobacterium tuberculosis or with anatomical pathological findings gave a picture in terms of which the phenomena associated with the clinical entity, consumption, could be collected and organized. This organization involved the binding of the phenomena of the syndrome to the etiological matrix of pathophysiological laws. The temptation was, though, as has been illustrated, to reify the matrix as an *ens morbi* – to, as Virchow remarked, treat disease as "a real substance (*ens*)."[21]

Similarly, the realist ontological theories involved a commitment to disease types. But reality was closer to that of the physiological theorists in being etiologically open – disease entities did not prove to have unique etiologies. As Virchow indicated, known pathogens can exist in hosts without pathology.[22] Disease causality is equivocal. Diseases are, as the physiological nosologists stressed, much more complex than the easy simplicity implicit in many ontological nosologies. Diseases are, in fact, not only multifactorial, but multidimensional, involving genetic, physiological, psychological, and sociological components. The presence of these various components does not merely entail a superimposition of modifying variables upon basic disease structures. Rather, it implies that diseases have a basically relational, not a subject (i.e., substance)-predicate (accident) nature. That is, there is not necessarily a *bearer* for every disease, a substrate for each type of disease.

This view of disease emerges from consideration of the complex of etiological structures involved in modern "disease entities." Diseases such as asthma, cancer, coronary artery disease, etc., are as much psychological as pathophysiological in that the likelihood of such illness is closely bound to experienced stress and the availability of support for the person stressed.[23] They are thus sociological as well. The result is a multidimensional concept of disease with each dimension – genetic, infectious, metabolic, psychological and social – containing a nexus of causes bound by their appropriate, usually different, nomological structures. The multiple factors in such well-established diseases as coronary artery disease suggest that the disease could be alternatively construed as a genetic, metabolic, anatomic, psychological, or sociological disease, depending on whether one was a geneticist, an internist, a surgeon, a psychiatrist, or a public health official. The construal would depend upon the particular scientist's appraisal of which etiological variables were most amenable to his manipulations. For example, the public health official may decide that the basic variables in coronary artery disease are elements of a lifestyle which includes little exercise, overeating and cigarette smoking. He may then address these social variables and consider such diseases to be, as Stewart Wolf suggested, ways of life.[24]

This shift in nosology is back to a "Hippocratic" notion of disease in the sense of a "physiological" or contextual concept. The Greeks have been criticized for producing "many separate disease entities, but never [arriving] at the concept of specific etiology."[25] In a sense, the criticism is justified. The Greeks described syndromes, but did not succeed in providing nomological substructures for diseases so that reliable explanation, prediction, and control of reality (i.e., diagnosis, prognosis, and therapy) were possible. The ontological nosologies can somewhat sum-

marily be described as a response to this shortcoming; they were a move
to a further level of abstraction allowing the signs and symptoms clustered
in a disease to be understood as the appearance of either a disease-thing
(an *ens morbi*), or a specific disease pattern. But the actual complexity of
diseases suggests that though Hippocratic medicine failed to advance
successfully a general account of the syndromes it described, it was properly cautious in not accepting theories of specific etiologies.

Conclusions in this area are at best tentative. But the ever more frequent
epidemiological studies of disease, such as the Framingham study of
cardiovascular disease, indicate a pattern-pattern analysis within which
the pattern of signs and symptoms clustering in a syndrome is bound
to a pattern of causal variables.[26] Appearance is bound to a nomological
substructure, with both levels having a fairly open-ended character.
Moreover, this open-endedness is controlled by pragmatic interests
which can bring the disease to a genetic, metabolic, psychological, or
social focus, etc., depending on which variables are to be manipulated.
Such focusing, though, does not require a reduction of other variables.
That is, the genetic variables in heart disease are not reduced to occult
sociological variables when one focuses on the elements of lifestyle central
to treating heart disease. They are rather treated as mediate variables
placed in relation to a model which has sociological variables and correlations as its central structure. One abandons ontological hypostatization of disease and nosological realism, and construes the reality of disease as a conceptual nexus posited for understanding the world of appearance.

Where does this put us with regard to models of disease? The adoption
of either a medical or psychological model is a pragmatic choice to focus
on a particular cluster of variables and their correlations in order to make
certain explanatory, predictive, and controlling maneuvers. But, to isolate
these distinguishable dimensions of diseases, is to separate that which is
distinguishable but of one fabric. The question of the correct model is
either a pragmatic question or a misunderstanding. All diseases can be
construed as both medical and psychological; only a confirmed Cartesian
would hold that the models are totally separable, while only a monist
would hold that they are not distinguishable. To assert that there is not
a somatic substrate for psychological events is to assert that psychological
life takes place nowhere in this world, that it is the enterprise of an at

least partially nonembodied spirit. If human experience and action is to be integrated in and for this world, it must occur somewhere in this world. On the other hand, those psychological generalizations which coordinate mental events in terms of drives and inclinations are distinguishable as such from models free of such intentional predicates and generalizations.[27] If mind and body are not two substances but two distinguishable levels of human significance, concerning which generalizations of different characters can be made, as indeed does appear to be the case, then medical and psychological models of disease should be complementary, not competitive. They should complete what would otherwise be one-sided assertions concerning a particular model.

The concept of disease is an attempt to correlate constellations of signs and symptoms for the purposes of explanation, prediction, and control. Pitfalls exist, such as the temptation to reify diseases or to treat diseases as rigid, specific types with unique etiologies. Diseases involve patterns of causes correlated with clusters of signs and symptoms which constitute the illnesses at hand. Disease models, as nomological patterns, are interpretable in larger patterns in a way not readily apparent if those models modeled disease things or independent disease types. Thus, one can concomitantly have a medical and a psychological account of the etiology of coronary artery disease, which are not incompatible, but mutually completing. That is, the models are not modeling two different things, nor are they two models of the same thing – a disease thing. Rather, they are two modes of correlating variables intrinsic and extrinsic to an ill person. The variables are chosen for the purpose of speaking about and altering that illness. They are relationships structured for particular diagnostic, prognostic, and therapeutic goals, and are based on distinctions between psychological and physical phenomena, between psychological and physical predicates. The issue of the medical versus the psychological model of disease is thus a rehearsal of the mind-body problem. It is bound in part to the claim that diseases are things, with the consequent problem whether a disease thing can have a psychological reality. The identification of objectivity with physical descriptions suggests that if diseases are things, then they have nothing to do with the subjective world of values and social relations often imported into concepts of mental illness. Under such a view, mental illnesses could be diseases only if they described the malfunctioning of a mental thing (e.g., a *res cogitans*).

Criticisms of medical model accounts of mental illness are heterogeneous.[28] The critique given by Szasz turns in part on the notion that
problems in living are diseases only if they have a physical basis and that,
moreover, true diseases are value-free.[29] Under this view, all diseases are
medical diseases. As a consequence, in such a view psychiatry becomes
a moral enterprise where blame and praise (i.e., responsibility for one's
own actions) are more appropriate than treatment, prognosis, and therapy.[30] The myth of mental illness is the myth that the autonomy of mental
life is intrinsically undermined by disease. Disease in this account is
identical with a form of reductionism which hides social judgment under
claims to scientific objectivity. The assumption is that medicine deals
with things and that psychology deals with broader issues such as social
development.[31] But if diseases are means for coordinating phenomena
for the purposes of prognosis, diagnosis, and therapy, then the issues can
be reformulated not only to allow for the coordination of mental phenomena in diseases, but for the intrusion of values into medical models
of diseases as well. To talk of diseases, and an intrinsic role for values
in medical diseases, is to abandon ontological nosological analyses of
disease and replace them with a contextual view closer to the more open-
ended physiological nosologies of the past.

Diseases such as cancer, tuberculosis, and schizophrenia thus exist, but
as patterns of explanation, not as things in themselves or as eidetic types
of phenomena. Owsei Temkin comes close to such an appreciation of the
plastic nature of the concept of disease. "The question: does disease exist
or are there only sick persons? is an abstract one and, in that form, does
not allow a meaningful answer. Disease is not simply either the one or
the other. Rather, it must be thought of as the circumstance requires.
The circumstances are represented by the patient, the physician, the
public health man, the medical scientist, the pharmaceutical industry,
society at large, and last but not least, the disease itself."[32] But the disease in itself is in the end the disease as it exists for us who both experience illness and explain it. Disease as an explanatory account is bound
to the circumstances of that account. In short, explanatory accounts are
not things; things are what explanatory accounts explain and disease is
a mode for explaining things – in particular, ill humans.

The portrayal of particular diseases involves pragmatic judgments
which ontological nosologies reified or stereotyped. C. S. Peirce argued

that "*In order to ascertain the meaning of an intellectual conception one should consider what practical consequences might conceivably result by necessity from the truth of that conception; and the sum of these consequences will constitute the entire meaning of the conception.*"[33] That is, evaluation enters into the enterprise of medical explanation because accounts of disease are immediately focused on controlling and eliminating circumstances judged to be a disvalue. The judgments are in no sense pragmatically neutral. Choosing to call a set of phenomena a disease involves a commitment to medical intervention, the assignment of the sick role, and the enlistment in action of health professionals. To call alcoholism, homosexuality, presbyopia, or minor hookworm infestation diseases, involves judgments closely bound to value judgments. Granted, there is a spectrum from broken limbs to color blindness along which interest in construing a constellation of phenomena as a disease varies. The pain and discomfort of either a broken limb or a schizophrenic break invite immediate medical aid, while issues of color blindness or dissocial behavior lie at the other end of the spectrum. But all along the spectrum, the concept of disease is as much a mode of evaluating as explaining reality.

Commitment to the concept of disease presupposes that there are phenomena physical and mental which can be correlated with events of pain and suffering, so that their patterns can be explained, their courses predicted, and their outcomes influenced favorably. Further, the pain and suffering cannot be the immediate outcome of circumstances which are directly the subject of free choice. They must result from psychological or physiological laws; that is, they must be open to statement in the form of laws, not moral rules. Medicine is the application of scientific, not moral generalizations. Thus, involutional melancholia, duodenal ulcer, and pneumothorax due to gunshot count as diseases, while ignorance, greed, and political violence do not, insofar as the ignorant are capable of learning, the greedy of virtue, and the violent of pacific action. Thus, mental deficiency, kleptomania, and paranoid reactions do count as diseases.

Of course, a broad concept of diseases including both mental and physical models opens one to greater influence by social values. Yet, social judgments are involved in not considering such events as childbirth to be diseases, even though they are associated with considerable morbidity and in fact mortality. Socially desirable goals help draw the lines.[34] The same is true with regard to aging and what then counts as disease and health. The acceptable physical state of an 80-year-old would be disease for a 20-year-old. Yet, will that always be the case as more can be effected through geriatrics? Diseases are, as Lester King has indicated, patterns which we structure according to our expectations.[35]

This is not to argue that confusion between medical and moral issues is not possible or that such confusion is unlikely. Quite the contrary, such confusion is very likely, given the nature of disease. The concept of disease has fuzzy borders with moral concepts. In the 19th century, for example, masturbation was considered to be primarily a physical disease in the same sense that someone now may hold that coronary artery disease is a physical disease though certain forms of stress may be necessary conditions for the disease.[36] More alarming examples exist, such as diseases developed to modify political behavior, such as drapetomania, the running away of slaves.[37] Szasz interprets such concepts as a reductio of the use of disease models in psychiatry, rather than as a caveat with regard to confusing compulsive and free action.[38] One of Cartwright's contemporary critics was more to the point in his remark that "if a strong desire to do what is wrong be a disease, the violation of any one of the Ten Commandments will furnish us with a new [disease]..."[39] In short, Cartwright's explanation failed because treating runaway slaves as free agents was a better account than treating them as subjects of fugue states.

Of course, one can still use medical force for political ends. Hookworms can be treated to eliminate anemia and thus make citizens more alert, or conceivably, citizens could be drugged into lethargy. The difference is that the first, unlike the second, is focused as well on the autonomy of the individual, his health. The accent of medicine upon liberating individuals from the hindrances of otherwise uncontrollable psychological and physiological forces, is the focus of the concept of health. The concept of health helps define the concept of disease by providing the telos for the medical enterprise. Disease concepts are, as has been argued, pragmatic concepts whose truth is found in action directed to the elimination of illness and toward the establishment of health.

III. HEALTH AND DISEASE

Models and accounts of disease are necessarily varied – they focus on the varied, particular limitations to human life. Health, though, represents a direction common to all the continua from particular diseases to well-being. If health is a state of freedom from the compulsion of psychological and physiological forces, there is a common leitmotif in the treatment of either schizophrenia or congestive heart failure – namely, the focus on securing the autonomy of the individual from a particular class of restrictions.[40] The unity of models of diseases is found more in the concept of health than in the concept of disease. Health is the common way away from the many ways of disease.

Thus, while the concept of disease is both an evaluatory and explanatory concept, health as a positive concept is more a regulative ideal. This may account in part for the difficulties surrounding attempts to define health operationally, though operational definitions of freedom from particular diseases are more available. Finally, to stress the non-moral character of the concept of health, a reminder from Freud is appropriate: treatment "does not set out to make pathological reactions impossible, but to give the patient... freedom to decide one way or the other."[41] Medicine, whether in the case of medical or psychological models of disease, is not an enterprise of applied ethics,[42] though values influence what limitations on human actions will be considered significant, and at times lead to confusing vices with the compulsions of nature.

In conclusion, health and disease are not symmetrical concepts, nor are they things, though important confusions have arisen from conceiving of them as such. Rather, the concept of disease is a mode of analyzing certain phenomena for the purposes of diagnosis, prognosis, and therapy. The concept is in one respect pragmatic, and in many respects influenced by issues of value. Particular diseases border on questions of moral and political significance. And, while there are many diseases, there is in a sense only one health – a regulative ideal of autonomy directing the physician to the patient as person, the sufferer of the illness, and the reason for all the concern and activity.

University of Texas Medical Branch,
Galveston, Texas

NOTES

[1] Constitution of the World Health Organization (preamble). *The First Ten Years of the World Health Organization* (Geneva: W.H.O., 1958).

[2] Alvan R. Feinstein, *Clinical Judgment* (Baltimore: The Williams and Wilkins Company, 1967), pp. 145–148.

[3] Miriam Siegler and Humphry Osmond, "The 'Sick Role' Revisited," *The Hastings Center Studies* 1 (1973), 41–58.

[4] H. Tristram Engelhardt, Jr., "Explanatory Models in Medicine: Facts, Theories, and Values," *Texas Reports on Biology and Medicine* 32 (Spring 1974), 225–39.

[5] Henry E. Sigerist, *Man and Medicine*, trans. by Margaret G. Boise (New York: W. W. Norton & Co., 1932), pp. 105–106.

[6] Walter Pagel, *Paracelsus* (Basel, Switzerland: S. Karger, 1958), p. 137.

[7] Walter Pagel, *The Religious and Philosophical Aspects of van Helmont's Science and Medicine* (Baltimore: The Johns Hopkins Press, 1944), pp. 39, 41.

[8] Walter Pagel and Marianne Winder, "Harvey and the 'Modern' Concept of Disease," *Bulletin of the History of Medicine* 42 (1968), 496–509.

[9] Knud Faber, *Nosography in Modern Internal Medicine* (New York: Paul B. Hoeber, 1923), p. 98; see also pp. 108–109.

[10] Rudolf Virchow, *Hundert Jahre allgemeiner Pathologie* (Berlin: Verlag von August Hirschwald, 1895), p. 22; English trans. by Lelland J. Rather, *Disease, Life, and Man: Selected Essays by Rudolf Virchow* (Stanford: Stanford University Press, 1958), p. 192.

[11] *Ibid.*

[12] Virchow, *Hundert Jahre allgemeiner Pathologie*, p. 23; *Disease, Life, and Man*, p. 192.

[13] Francois Boissier de Sauvages de la Croix, *Nosologia Methodica, Sistens Morborum Classica. Juxta Sydenhami menten et Botanicorum Ordinem* (Amsterdam: Fratrum de Tournes, 1768).

[14] Philippe Pinel, *Nosographie philosophique, ou la méthode de l'analyse appliqué à la médecine* (Paris: Brosseau, 1798).

[15] F. J. V. Broussais, *Examen des Doctrines Medicales et des Systemes de Nosologie*, Vol. 2 (Paris: Mequignon-Marvis, 1821), p. 646.

[16] Peter H. Niebyl, "Sennert, van Helmont, and Medical Ontology," *Bulletin of the History of Medicine* 45 (1971), 118.

[17] Faber, *Nosography*, p. 95.

[18] Henry Cohen, *Concepts of Medicine*, ed. by Brandon Lush (Oxford: Pergamon Press, 1960), p. 160.

[19] Carl A. Wunderlich, "Einleitung," *Archiv für physiologische Heilkunde* 1 (1842), v.n.; ix.

[20] Ernst Romberg, *Lehrbuch der Krankheiten des Herzens und der Blutgefässe*, 2nd ed. (Stuttgart: Verlag von Ferdinand Enke, 1909), p. 4.

[21] Virchow, *Hundert Jahre allgemeiner Pathologie*, p. 22; *Disease, Life, and Man*, p. 191.

[22] *Ibid.*, p. 38.

[23] Thomas H. Holmes and Richard H. Rahe, "The Social Readjustment Rating Scale," *Journal of Psychosomatic Research* 11 (1967), 213–218.

[24] Stewart Wolf, "Disease As a Way of Life: Neural Integration in Systematic Pathology," *Perspectives in Biology and Medicine* 4 (Spring 1961), 288–305.

[25] Robert P. Hudson, "The Concept of Disease," *Annals of Internal Medicine* 65 (September 1966), 598.

[26] U. S. National Heart Institute, *The Framingham Study: An Epidemiological Investigation of Cardiovascular Disease*. Section 1, p. 1b–6, U. S. Government Printing Office, Washington, D. C., 1968.

[27] H. Tristram Engelhardt, Jr., *Mind-Body: A Categorial Relation* (The Hague: Martinus Nijhoff, 1973), pp. 148–161.

[28] Ruth Macklin, "Mental Health and Mental Illness: Some Problems of Definition and

Concept Formation," *Philosophy of Science* **39** (September 1972), 341–365, and "The Medical Model in Psychoanalysis and Psychotherapy," *Comprehensive Psychiatry* **14** (January/February 1973), 49–69.

[29] Thomas S. Szasz, "The Myth of Mental Illness," *The American Psychologist* **15** (February 1960), 113–118.

[30] Thomas S. Szasz, *The Ethics of Psychoanalysis* (New York/London: Basic Books, Inc., Publishers, 1965).

[31] George W. Albee, "Emerging Concepts of Mental Illness and Models of Treatment: The Psychological Point of View," *American Journal of Psychiatry* **125** (January 1969), 870–876.

[32] Owsei Temkin, "The Scientific Approach to Disease: Specific Entity and Individual Sickness," in *Scientific Change*, ed. by A. C. Crombie (London: Heinemann, 1961), pp. 629–647.

[33] C. S. Peirce, *Collected Papers*, ed. by Charles Hartshorne and Paul Weis (Cambridge, Mass.: Belknap Press, 1965), 5.9.

[34] Lester S. King, "What Is Disease?," *Philosophy of Science* **21** (July 1954), 197.

[35] *Ibid.*

[36] H. Tristram Engelhardt, Jr., "The Disease of Masturbation: Values and the Concept of Disease," *Bulletin of the History of Medicine* **48** (Summer 1974), 234–248.

[37] Samuel A. Cartwright, "Report on the Diseases and Physical Peculiarities of the Negro Race," *The New Orleans Medical and Surgical Journal* **7** (May 1851), 707–709.

[38] Thomas S. Szasz, "The Sane Slave," *American Journal of Psychotherapy* **25** (April 1971), 228–239.

[39] James T. Smith, "Review of Dr. Cartwright's Report on the Diseases and Physical Peculiarities of the Negro Race," *The New Orleans Medical and Surgical Journal* **8** (September 1851), 233.

[40] Much is packed in and hidden away in the notion of a "particular class of restrictions" which has only been sketched in part above.

[41] Sigmund Freud, *The Standard Edition of the Complete Psychological Works of Sigmund Freud*, ed. and trans. by James Strachey (London: Hogarth Press and The Institute of Psycho-Analysis, 1961), Vol. 19, *The Ego and the Id and Other Works*, p. 50, n.1.

[42] H. Tristram Engelhardt, Jr., "Psychotherapy as Meta-ethics," *Psychiatry* **36** (November 1973), 440–445.

LORETTA KOPELMAN

ON DISEASE: THEORIES OF DISEASE AND THE ASCRIPTION OF DISEASE: COMMENTS ON "THE CONCEPTS OF HEALTH AND DISEASE"*

Dr. Engelhardt has argued that the concept of disease is ambiguous because it has both an explanatory and an evaluative role.[1] I would like to examine this. To do so I will begin by distinguishing between theories of disease on the one hand and the ascription of disease on the other.

By theories of disease I mean medical-scientific theories which include generalizations about certain clusters or sequences of phenomena which are seen as, called, or regarded by a society or medical experts as pathological. Medical-scientific theories are distinguishable from other scientific theories in that the patterns or sequences studied have to do with the structure and function of human organisms. Like other scientific theories they are systematically related sets of statements including some generalizations that are law-like and empirically testable. Framing and testing scientific theories involve many judgments and evaluations. However, the theories themselves are distinguishable from normative judgments about whether the clusters, sequences, or patterns studied are of value or disvalue in the sense that they are viewed as disease or as health. Dr. Lester King writes, "All medical science studies facets of behavior under a wide variation in conditions. Many of these variations we call disease. But the grounds for calling them disease are not any part of the study."[2]

The ascription of disease is different from medical theories about disease. The ascription of disease is the seeing of something as a disease and naming of it as a disease. It is the viewing and calling of a sequence, cluster, or pattern of phenomena as a disease. Disease may be ascribed to individuals differently in different societies. For example, in some societies homosexuality is considered a disease, but in others it is not.[3] Also, disease may be ascribed to individuals whether or not their symptoms are encompassed by a medical theory, or even if there is no theory of disease which explains their symptoms. Moreover, it is possible for someone to have a disease even though the person is not recognized as having a disease in his own society. For example, suppose a culture did

H. T. Engelhardt, Jr. and S. F. Spicker (eds.), Evaluation and Explanation in the Biomedical Sciences, 143–150.
All Rights Reserved. Copyright © 1975 by D. Reidel Publishing Company, Dordrecht-Holland.

not ever ascribe or call something a disease unless or until the person had pain, discomfort, or the impairment of function. In such a society, based upon views about how to ascribe disease to individuals, a person with symptomless carcinoma of the lung would not be regarded as having a disease. Based upon our modern medical theories and our views about how to ascribe disease to individuals, however, it would be improper for us to regard someone with symptomless carcinoma of the lung as not having a disease.

Sickle cell trait offers a very clear illustration of the difference between theories about disease and the ascription of disease in different cultures. This is a condition which occurs in black individuals who inherit an abnormal hemoglobin molecule in their red blood cells. These people have no symptoms at all except under unusual circumstances when oxygen is scarce. For instance, at very high altitudes they are prone to serious complications such as vascular accidents (strokes, etc.).[4] In our culture individuals with sickle cell trait may be considered to have a disease, for they may not survive sudden accidental depressurization of an airplane; accordingly, they are forbidden from being on flight crews.

In contrast, in African communities when air travel did not exist and where malaria was common, having sickle cell trait was not a disvalue at all. Individuals with sickle cell trait have increased immunity from malaria when compared with other people. In such African communities the same condition was of positive value, and would not have been considered a disease state even if such individuals could have been easily identified. The theory about sickle cell trait is the same, but the regarding of it as a disease is different, depending upon whether we are talking about our culture or an African community where malaria but not air travel was common.[5]

Our choice of theories will influence what we see as a disease, and what we see as a disease will influence our theories about disease. Current discussions in psychopathology illustrate that we can have genuine disputes and/or doubts about the truth or adequacy of theories of disease, and about what ought to be regarded as a disease.[6]

"Disease" is ambiguous. It may refer to theories of disease which, like other scientific theories, are empirically testable, or it may refer to the ascription of disease. Dr. Engelhardt puts together under the title "the concept of disease" these two different and distinguishable notions. If we keep the distinction between theories of disease and the ascription of disease before us, then we shall, I think, be able to clarify certain apparent ambiguities in his statements about "the concept of disease." I will argue that some of Dr. Engelhardt's statements on "the concept of disease" apply to one but not the other of these two notions.

In this regard, I would like to discuss five important points which Dr. Engelhardt uses in his analysis of the concept of disease. They are: Point I, "The concept of disease is an attempt to correlate constellations of signs and symptoms for the purposes of explanation, prediction, and control", i.e., diagnosis, prognosis and therapy. (126, 135, 136) Point II, Diseases are not voluntary. (138) Point III, The designation of something as a disease indicates that it is regarded as a disvalue. (127, 137) Point IV, "The concept of disease has fuzzy borders with moral concepts." (138) Finally, Point V, "Choosing to call a set of phenomena a disease involves a commitment to medical intervention, the assignment of a sick role, and the enlistment in action of health professionals." (137)

Let us consider each of these five points which Dr. Engelhardt makes about the concept of disease, in relation to the distinction which we made between theories of disease and the ascription of disease. The first point that Dr. Engelhardt states is that the concept of disease involves an attempt to correlate certain phenomena for the purpose of explanation, prediction, and control of reality, i.e., diagnosis, prognosis, and therapy. Medical theories, like other scientific theories, attempt to explain and predict, and are used to control reality. Dr. Engelhardt's first point, then, certainly applies to theories of disease. Let us see whether it also applies to ascribing disease to individuals.

In ascribing disease we are not necessarily framing scientific theories or explaining anything. Consider the following imaginary case. Suppose that someone had a fever, malaise, and green spots. There was nothing wrong with the lighting or our eyesight, it was not a sham, and no explanations (such as a parasitic infection) could be found. Suppose that we were unable to explain the green spots in a way compatible with present theories. Still we would ascribe illness and disease to the individual even though we had no theory to account for this disease. Suppose that we gave up all hopes of finding an explanation for it based upon current understanding. We would still, I think, call it a disease. Therefore, it does not seem to me that point I as it stands is a necessary condition for the calling of something a disease. That is, we may call something a disease even though in doing so we have no goal of explanation and prediction. Point I, then, is true of disease in the sense of theories of disease, but not necessarily true about the ascription of disease.

Let us now consider a second point Dr. Engelhardt makes about the concept of disease. Point II states that diseases are not voluntary. If we found that an individual could control his symptoms and signs at will, then we would not call his condition a disease. Its involuntary nature, then, does seem to be a necessary condition for the ascription of disease. On the other hand, one can have scientific theories about any experience or phenomena, whether or not they are under the voluntary control of persons. Repeatable phenomena can be studied and theories developed to try to explain them, whether or not they are under voluntary control. If something is called a disease then it is believed to be involuntary.

A third point Dr. Engelhardt makes with regard to the concept of disease is that the designation of something as a disease indicates that it is regarded as a disvalue. Scientific theories in themselves do not contain such value judgments about the relative desirability or value of the phenomena studied. Of course, certain constellations of symptoms and signs are selected for study because they are judged to be disvalues. For instance, a process that causes suffering or shortens life should and will receive priority attention by medical scientists as compared with an innocuous curiosity. Therefore, point III does not apply to medical-scientific theories. Let us now see whether this point applies to disease in the sense of ascribing disease.

If point III is meant to be a necessary condition for ascribing disease or justifiably calling something a disease, then it is not possible both to justifiably regard something as a disease and also to value it. Dr. Engelhardt suggests that viewing something as a disvalue is a necessary condition for calling something a disease, but the suggestion is not defended. He writes, "... [the concept of disease] indicates a state of affairs as undesirable and to be overcome," (127) and "To call [something] a disease involves judgments closely bound to value judgments." (137) Whether he fully commits himself or not, it seems to me to be correct, but not self-evident, to say that regarding something as a disvalue is a necessary condition for ascribing disease. The sickle cell trait example that we discussed earlier illustrates that we do not call something a disease unless it is a disvalue. Where malaria is common and air travel is not, then this trait is a positive value and not regarded as a disease. Where malaria is uncommon and air travel is common then it is a disvalue and is considered a disease.

Let us consider another example. In the eighteenth century smallpox was a common and dreaded disease. Edward Jenner discovered that persons who had contracted the mild disease cowpox did not get smallpox. He developed a method for infecting persons with cowpox in order to protect them from smallpox.[7] While it would be a mistake to say that the cowpox in itself was valued, its effects were valued, and the disease was preferred as, by far, the lesser of two evils. More recently a vaccine against smallpox was made using vaccinia virus, a virus closely related to the one that causes cowpox. After a person is given a smallpox vaccine the vaccinated site becomes swollen, red and tender and the person may develop fever and malaise.[8] Yet his symptoms and signs, which are virtually identical to those of cowpox, are no longer considered to be indicative of a disease state. They are caused by the vaccination "taking," and this is of value because the individual will then have immunity to smallpox. The same signs and symptoms may represent a process which is of value or of disvalue. Only if the process is seen as being a disvalue do we say the person has a disease. The pains of childbirth do not indicate a disease process.

I do not believe that it is a counter-example to this to show that some of the believed side-effects of a disease have been valued. There was a popular belief or superstition from ancient times until the Renaissance that epilepsy was associated with prophetic powers. In this case, a supposed effect, or phenomenon believed to be correlated with the disease, was valued. However, Owsei Temkin argues that epilepsy was not valued by physicians or the general public in itself, and that it was called the divine sickness in ancient Greece because it was believed that only divine intervention could cure it.[9] In the nineteenth century some of the effects of tuberculosis were idealized by poets and artists of the period. It was supposed to increase or intensify intellectual or artistic performance, and the ravages of the disease were even regarded as beautiful.[10] However, here too it was not the disease in itself which was valued. To the contrary, it was dreaded. Rather, it was some supposed side effects or phenomena believed to be correlated with it that were valued by a small part of the population.

Point III, then, does apply to disease in the sense of ascription of disease, for it does seem that if we regard something as a disease then we regard it as a disvalue. However, point III does not apply to disease in the sense of theories about disease, because theories about disease, like other scientific theories, do not, in themselves, contain value judgments about the undesirability, danger, or ugliness of the phenomena theorized about.

Point IV is, "The concept of disease has fuzzy borders with moral concepts." (138) This is true of the ascription of disease. For example, homosexuality may have been called a disease because it was the object of disapproval. In contrast, theories of disease, which are empirical and contingent, should not be so difficult to keep separate from moral categories.

Point V is "Choosing to call a set of phenomena a disease involves a commitment to medical intervention, the assignment of the sick role, and the enlistment in action of health professionals." (137) This comment does not refer to medical theories in themselves, because scientific theories do not commit us to such action. Neither does it necessarily refer to ascription of disease in itself. It is possible to ascribe disease to a person and not take *all* of these actions. A slight cold or a superficial infection, while they may deserve some attention, do not warrant enlistment of health professionals. These actions may be associated with the ascription of disease, but neither medical theories nor ascriptions of disease in themselves necessitate that action to eliminate illness and restore health be taken. However, this is certainly the usual goal or purpose of those involved in the ascription and study of disease.

In summary I have argued that Dr. Engelhardt puts together under the heading of "the concept of disease" two different and distinguishable notions, namely, theories of disease and the ascription of disease. I have tried to show that many of the comments he makes about "the concept of disease" apply to one, but not the other of these two notions. I would like to point out another example of the usefulness of distinguishing between the ascription of disease and theories of disease.

In the body of his paper Dr. Engelhardt criticizes ontological theories about disease for not making room for values. He argues in favor of a view which he claims is closer to the more open-ended physiological theories of the past and present. He writes, "But if diseases are means for coordinating phenomena for the purposes of prognosis, diagnosis, and therapy, then the issues can be reformulated not only to allow for the coordination of mental phenomena in diseases, but for the intrusion of values into medical models of diseases as well. To talk of diseases, and an intrinsic role for values in medical diseases, is to abandon ontological nosological analyses of disease and replace them with a contextual view closer to the more open-ended physiological nosologies of the past." (136)

However, I believe that ontological theories, like the ancient forerunner of the physiological theories that disease is the malmixture of the humors, and like more modern physiological theories, are all intended primarily as empirically testable theories about sequences or clusters of phenomena regarded as disease states. It seems to me to be a mistake to criticize the ontological theories because they do not make room for the intrusion of values of the sort that we have been discussing, or to praise physiological theories because they do. That is, it seems to me to be a mistake to suggest that medical scientific theories in themselves properly contain value judgments of the sort which have been discussed as appropriate to the ascription of disease; namely, those judgments or evaluations which involve or border on moral, social or aesthetic norms. Were it the case that medical scientific theories in themselves properly contained judgments and evaluations of *this* sort, then they would be unlike other scientific theories. Dr. Engelhardt seems to suggest this and it seems to me to be a mistake. I think the problem here is the failure to fully distinguish between what I have called theories of disease and the ascription of disease.

In conclusion, I have tried to separate and distinguish between these two concepts: medical theories of disease and the ascription of disease. In practice, of course, they are very closely related. The same individuals, physicians, are experts both in ascribing disease to individuals, and also in proposing, evaluating, and knowing of medical theories about disease. Distinguishing between them for the purpose of analysis, however, is very important. Otherwise we can fall into the error of supposing that some scientific theories properly contain value judgments of the sort we have discussed or the error of supposing that in ascribing disease we always have as our goal framing theories, explanation, prediction, and control of reality.

University of Rochester School of Medicine, Rochester, New York

NOTES

* This paper was completed with the aid of the National Endowment for the Humanities, Younger Humanist Grant, 1973–1974.

[1] H. Tristram Engelhardt, "The Concepts of Health and Disease," in this volume, p. 125. Henceforth all references to this work will be included parenthetically in the text.

[2] Lester King, "What is Disease?," *Philosophy of Science* 21 (July 1954), 194.

[3] In our society in recent years, psychiatric theories about, e.g., the causes of homosexuality, have not changed as dramatically as the way that it is viewed. Formerly it was viewed by many psychiatrists as a disease, but this is no longer the case.

[4] M. M. Wintrobe, *Clinical Hematology*, 6th ed. (Philadelphia: Lea and Febig, 1967), pp. 690 and 705.

[5] There has been a controversy about whether sickle cell trait should be regarded as a disease. Such controversy further illustrates the central point of this paper, namely, the importance of distinguishing between the ascription of disease, and theories which include generalizations about phenomena regarded (sometimes controversially) as pathological. The ascription of disease, that is, the viewing and naming of certain phenomena such as sickle cell trait as a disease, may be related to moral, aesthetic, social or political considerations. However, this is different from the theories, norms and judgments which go into stating what sickle cell trait is, how it is caused, and what its effects are like. This distinction is not intended to be any overall separation of facts and values. It is not intended to be any sort of a defense of positivism, physicalism, or reductionism.

[6] See, for example, the discussion in *Psychopathology Today*, ed. by William S. Sahakian (Itasca, Illinois: F. E. Peacock, 1970). Of particular interest in this regard are articles by Thomas Szasz, "Repudiation of the Medical Model," pp. 47–53; D. A. Ausbel, "Medical or Disease Model," pp. 53–60; H. J. Eysenck, "Learning Theory Model," pp. 73–85; and Margaret Mead, "Socio-Cultural Model," pp. 85–100. Ruth Macklin has two excellent discussions of this: "The Medical Model in Psychoanalysis and Psychotherapy," *Comprehensive Psychiatry* 14 (Jan./Feb. 1973), 49–69, and "Mental Health and Mental Illness: Some Problems of Definition and Concept Formation," *Philosophy of Science* 39 (September 1972), 341–365. See also Marie Jahoda, *Current Concepts of Positive Mental Health* (New York/London: Basic Books, 1958), chap. II.

[7] S. Krugman and R. Ward, *Infectious Diseases of Children and Adults*, 5th ed. (St. Louis: C. V. Mosby Co., 1973), p. 277.

[8] *Ibid.*, p. 278.

[9] Owsei Temkin, *The Falling Sickness*, 2nd ed., rev. (Baltimore: The Johns Hopkins Press, 1971), p. 4, pp. 153–154.

[10] René and Jean Dubos, "Consumption and the Romantic Age," in *The White Plague* (Boston: Little, Brown and Co., 1952), chap. 5.

[11] Theories *in themselves* are meant to contrast to related but distinguishable notions such as what motivates us to form them, or what leads us to believe that they are useful or even dangerous.

BODY AND SELF:
PHENOMENOLOGICAL PERSPECTIVES

RICHARD M. ZANER

CONTEXT AND REFLEXIVITY:
THE GENEALOGY OF SELF

I

The project I have set myself is mammoth and unmanageable. To attempt it anyway is foolish, and not only for that reason. As William Golding's Jocelin reflects, "to think how the mind touches all things with law, yet deceives itself as easily as a child,"[1] so this project, seeking the inner *logos* of self's emergence, may well be too easily deceived. "Self," so readily characterizable in language by a substantivization of a reflexive (divested of the pronominative it qualifies), is, it may be, no substantive at all but an oddly fugitive reflexive presence, foredooming such inquisitive efforts as this to subtle but always rude failure. The vagaries of custom and habit seduce one to believe in the continuous subsistence of self, sometimes with marked passion; yet searches designed to ferret it out just as often end by entrapping themselves in their own belief and passion. It were wiser perhaps to leave these matters to those more accustomed to the regions of illusion and sly dexterity: the magical crafts.

Yet the self is a notion often invoked both by medicine generally and psychiatry in particular. If not presupposed explicitly, it lurks importantly in the background of thought about the patient as a person, as an individual, as a particular human. Notions of the self and the significance of the self structure the backdrop of many reflections concerning human deportment such as references to self-punishing behavior or obsessive self-reproach. The self is a notion centrally invoked in all talk about man, but most importantly so in medicine. Yet on the whole references to the self have remained naive and uncritical.

What is the "self"? Opinions on that have varied. There are the affirmers of substance, such as Augustine, Descartes, Leibniz, or contemporary theorists like Macmurray. There are the firmly implanted deniers: the early Sophists and skeptics, the redoubtable Hume, and contemporaries like Sartre. There are also those who argue the *tout naturel*: self is "wholly natural," and thus indistinguishable from brain-states, complex

H. T. Engelhardt, Jr. and S. F. Spicker (eds.), Evaluation and Explanation in the Biomedical Sciences, 153–174.
All Rights Reserved. Copyright © 1975 by D. Reidel Publishing Company, Dordrecht-Holland.

neurophysiology, and the like – one thinks of the atomists, Hobbes, and today of "identity" theorists such as Armstrong. And then there is an array of uncataloguables, bewitchers of the common tongue: from Herakleitos and his *Logos* forever saying dark things about opposites which are the same and the endless labyrinth which is the self, Pascal and his tender, tenacious reed with its logic of the heart, Kierkegaard and his linguistically acrobatic play on relations relating themselves to themselves, to present-day people such as Merleau-Ponty and Marcel trying to recapture the living unity of human being-at-the-world, or Paul Ricoeur who, denying immediacy of self to itself, tries to penetrate it hermeneutically, reading it through its often wispy traces and haunting symbols like a patient textual scholar bent on painful exegesis.

A quite considerable cluster of terms is to be found in such discussions, often used to indicate much the same reality: self, spirit, soul, psyche, subjectivity, subject, inner man, person; but also mind, consciousness, mental substance; as well as mental, psychic, subjective, personal, human, spiritual, and conscious life. Beyond these are some by now well-known philosophical expressions: ego, monad, transcendental unity of apperception, *Dasein, pour-soi, être-au-monde,* transcendental ego, and a few others. Other expressions also crop up, such as the common way of distinguishing among the material self, the empirical self, the social self, and the spiritual self; as well as the classical psychoanalytical ones: id, ego, superego, libido, sub- and unconscious, intrapsychic self, and the like. In fact, the more one reads in various literatures, the more bewildering becomes the linguistic and conceptual baggage. It seems an embarrassment, but hardly one of riches!

Such a profusion seems inevitably a confusion as well, and we are at a loss not only linguistically and conceptually but as regards the "things themselves." It takes but little effort to begin to appreciate the utter dismay some biomedical scientists, and some philosophers, have felt over the whole affair, ending by regarding all such discussions either as spurious, or as mere epiphenomenal fluff quite unnecessary in scientific and philosophic inquiries. There are problems and irritations enough, after all, simply trying to understand the incredible complexities of the body's many systems – not to mention the brain; to introduce such mentalistic vagaries seems merely to practice the deliberate craft of deception where clarity is already at a premium.

Here, as T. S. Eliot elsewhere notes with insight,

> Words strain,
> Crack and sometimes break, under the burden,
> Under the tension, slip, slide, perish,
> Decay with imprecision, will not stay in place,
> Will not stay still. Shrieking voices
> Scolding, mocking, or merely chattering,
> Always assail them.[2]

II

For all that, it is no mere idle curiosity that the many urgent questions people of all sorts ask about self – Who am I? What and why am I? What do I live for? What is my destiny? and others – keep cropping up, will not stay silent, not only in spite of, but often just because of that very confusion and dismay. We are, it seems, at an historical time when ways of thinking about human life, values, persons – the self – are not readily available, or not on hand at all for us. Yet, just these issues seem most pressingly critical for us, as is clear from even casual acquaintance with disturbed persons. Are "disturbed persons" nevertheless persons? Are there disturbances which can cancel out or so grievously affect the "person" as to imply the loss of "self"? Are fetuses *persons*? Are they *selves*? Or is being a self (or person for that matter) merely potential in such a case? And even if the latter, what are we to say of seriously deformed newborns? Are they, even potentially, persons or selves? As is patently clear, medicine raises the question of "self" in a dramatically pressing way.

What, indeed, are we to think, and in which ways must our thoughts proceed, when we are called upon to think about the "self"? Expressed differently, thinking about, coming to understand and ultimately being able to account for the specifically human is in the end possible solely to the extent that we can comprehend that most profoundly subtle and elusive phenomenon of "self." The focus of medical science and practice must in some way come to rest just here – whether the concern be for the organism or for mental life, it is always ultimately "my" or "your" body and life which is at stake, and that cannot be avoided or evaded, on pain of losing the very point at issue. All matters human are expressions of self and of interchanges among selves. Thus every discipline focusing on "the human" and "the human world" – social and behavioral science, history, theology, as well as the biomedical sciences, medical practice, and legal and educational concern – finally depends for its sense and justifiability on the elucidation of the phenomenon of "self."

But understanding and accounting for self requires that our thinking shape itself in accord with what self is; our thinking of necessity must be accurate, fair and adequate to what we want to think about and thereby account for. I know of no other way to gain this access to the phenomenon, than to let it stand out for itself and become evident on its own. I want now to try and allow this to happen.

III

Let the following illustration carry the main burden of this descriptive explication. For convenience, the illustration must be kept brief, relying heavily on the common stock of knowledge to flesh it out.

Driving my car to work one day, I come to an intersection whose traffic lights are not working. Traffic is heavy, and being already late for work, I quickly check to see if there is any way to get around the snarl of vehicles; I look for side streets, openings in the traffic, anxiously check behind me, spot an opening leading to a side street, cut into it, and am off sailing around the blockage.

All very common, even humdrum, and one is free to imagine any of a range of consequences – from happy end to a small disturbance, an upset which leads to ulcers, or even the "Death of Ivan Ilych." So, too, is one free to imagine other variations, and variations on the variations.

IV

Even though limited, such examples do serve to allow certain common things to stand out. In the first place, there is an interruption in the usual taken-for-granted course of affairs: something happens as an obstacle, in view of which I am obliged to pay attention, become alert; I cannot simply proceed to drive in the usual way, at least not without obvious consequences. The interruption signifies that I must "stop and think": "stop" doing what I had been doing more or less automatically, and "think" to see if I can get out of the situation or resolve the interruption, so that I can again be on my way.

These considerations allow something else to become salient: such upsets or obstacles have special ways of *referring*. (1) The traffic jam refers *to me*; *I* am blocked, and *I* am obliged to "do something." (2) It refers, more specifically, to *my prevailing concerns*, to my "what I am doing" and "where I am going," and is more or less serious for me strictly in view of my present concerns and actions. If I am merely on a casual drive or trying to get to work when I am late, or urgently trying to get my injured child to the hospital, the way in which the traffic jam is "upsetting" becomes quite obvious: it is "bothersome," "a damned nuisance," an "urgent crisis," etc. The interruption thus gets its specific sense and emotive textures from these referrals. But, as well, (3) the upset includes multiple references to the concrete circumstances: the other cars, pedestrians, the specific streets, the time of day, the multiple recipes and mores of driving, the laws governing traffic and the typical things to do when such as traffic lights don't work; and still further horizons of references to the surrounding world of city life, references to "the times," the world at large – everything, in short, going to constitute this scene as that which it is. Obviously, these "referrals" mutually determine and delimit each other, mutually interrelate in complex ways, and co-define their various meanings: what I am here and now ("busy," "late," "irritated," "liable to quick anger," etc.), what my concerns are ("going to work," "going to the hospital," etc.), and what the scene itself is ("too heavily trafficked," "obstacle," "crisis," etc.). I know, of course, that others on the scene are likewise networks for these cross-referrals, that the scene may be less or more urgent for them, and so on. And, in obvious as well as subtle ways I must reckon with these sometimes conflicting networks (another person may move into the opening in traffic, I must watch out to avoid a collision, watch for pedestrians, try not to break the law, at least not too much, etc.).

Thus, another facet becomes salient: what I do at the time is itself functionally co-determined and reciprocally significant with reference to the actual circumstances, to where and when the upset occurs, to others present there, to what I have been accustomed to think and do before, and what I now want to do. In brief, the situation is a clear instance of an integral *context*. I shall have more to say of this momentarily. For now, what is meant can be put this way: to be confronted with such an upset is to be *contextured*, in the now relatively precise sense of being within a well-structured network of mutually determined referrals, within a complex of constituents the significances of which are found within this integral set of multiple interrelationships. The "scene" as a *whole* is precisely this system of mutually interlocking and multiple cross-referring constituents, and every item within it receives its particular significance and "place" in relation to every other one. Thus the range of what I am able to do or not to do in this setting is strictly defined within it. Actions such as getting out of the car, turning left or right, stopping and starting, blowing the horn, swearing, and the like, are possibilities having their strict locus within the context itself. Other possibilities – flying, swimming, playing ball, sleeping, etc. – are not at hand just because the context in question does not allow for them, or makes them more or less unlikely.

With this, another feature crops up: I am alerted, called on to "stop and think." That is, within a determinable range of variations, I must cease doing what I was doing, and do something else: I must "think" and then "do," and both require *effort*. What is this "thinking"? I am interrupted, and while remaining within the actual situation, must now in a way *distance myself* from it, or de-actualize it. In different terms, I look for "ways to get out"; my "thinking" is a seeking for "alternatives." Necessarily within the actual context of the interruption, my thinking has the significance of "thinking for *the possibly otherwise*," of "*possibilizing*," and depending on my own ingenuity and imaginativeness, I will have at hand for my thinking and doing a greater or lesser number of alternatives to choose from. However that may be, my "distancing while remaining within" the situation is a thinking for the possibly otherwise; it is a seeking, within the limits imposed by the context at hand, for ways to resolve the interruption, to rearrange the deranged course of affairs, in the light of the "possibles" I can come up with. This possibilizing thinking is, thus, a sort of leap into the "as-if" ("other possible ways" I can go), but it is a leap which goes on within and continually maintains before it the actual interruptive situation.

It is, we could say, *circum-stantial possibilizing*, a deactualization of the situation in order to actualize one or another alternative from those "as-ifs" I am obliged to consider so that I can be on my way again, continue my project, or even my life (depending upon what crops up as "at issue").

V

In many ways, I suspect, the course of any human life always involves just such situations: casual upsets, trivial disturbances, or more serious interruptions, and even impasses, and various re-establishments of the course of life. Human life is thus a network of mutually interlocking relationships and referrals constituting contextures within which alerted and effortful circum-stantial possibilizing go on designed to re-establish and/or re-direct prevailing concerns, in the ultimate interests of living. In these terms, any creature which is locked to the actual is a creature capable neither of history, culture, education, nor thinking – all these rest upon the ability to make that critical leap. Nor, it would seem, can one properly and sensibly speak, in such a creature, of "self."

VI

From such deliberations, it is possible now to specify somewhat better the sense of the phenomenon of "self." To be capable of "being interrupted" signifies a kind of *alertness*, a focal attending to something as blocking my path, and an *effortful* seeking to resolve it. But such alertness and effort, to be thus focused must of necessity also be *reflexive*: "to be alerted to" requires a recognition or awareness of something as disturbing or blocking one*self*, thus as referring to and soliciting that alertness. Similarly, "to make an effort" (to think, to seek for ways out, etc.) requires an alertness to the interrupting situation, and the exigency to "do something" in order to resolve it, and the necessity to distance oneself and possibilize – hence, is a reflexive awareness of oneself as "needing to do." "Self," thus, makes its appearance as a *reflexive alertness and effort*, articulated as circum-stantial possibilizing – strictly and necessarily within concrete contexts of concern and action. Furthermore, the "having to do something" in reference to some interruption involves some mode of *bodily enactment*: focusing my eyes, moving my limbs, turning my head, locomoting, and the like. Thus, the sense of "self" is closely interlaced with bodily movements, postures, gestures, attitudes, actions, and sensory fields whereby something is experienced. "Self" is always "embodied self," and "embodied self" is always contextually located within the circumstances or actional field of concerns.[3]

Finally, at this *still preliminary* stage, to the extent that self emerges in situations of contextual conflict, it is deeply textured with a sense of *import*, a concern for "having" to regain its pose or equilibrium. It is thus charged with a kind of emphasis or exigence. Alertness and effort have the significance of striving, desire, preference, and decision: in short, of volitional, emotive, and valuational effort responding to such conflicts. Such contextures, then, *call into question* what has hitherto not been in question. That is, they call out a searching for alternatives, and thus *bring into question* what has up to now been a relatively harmonious ongoing course of experience. This "call," whose force is to "call forth" responses – the "calling out" which "brings out" alertness and effort, as striving and concern for settling the now unsettled – is thus precisely the appearance of "self" as that specific "itself" which is "in question" and "called forth" as "having to do something" in order to maintain itself, re-establish itself, or re-direct itself.

VII

Illness is a special case of the circumstances which block the *self* and call one's alertness and effort to be borne upon the context, and is thus a uniquely illustrative case. In illness, one's *bodily enactment* fails. For instance, the man with angina tries to climb the stairs, yet his dull substernal pain prevents him. He is forced to rest. His project of climbing to the top of the stairs for the time is arrested. Instead, his thoughts are interrupted and replaced, to quote Heberden, with "a sense of strangling and anxiety."[4] Angina is not for him just an external force, or even just an internal process of his body, but an important element of the context which defines his life, and within which he defines himself.

When he first developed angina, what had hitherto *not* been called into question, now has been. Stairs take on a new proportion. New things become relevant – having an elevator installed in the house, walking up only so many steps or at a certain pace, placing a nitroglycerin tablet sublingually. His case is similar to mine in the traffic jam, yet here it has been his body itself which has called things into question, forced him to be alert, brought him to search out alternatives – to rest, or to take a nitroglycerin tablet and attempt to finish climbing the stairs.

Illness calls into question aspects of the human context, usually taken for granted. The self is forced to maintain itself and to redirect itself.[5] The self as a contextual achievement is precarious, and has its precarious nature suggestively presented in illness. Consider again, if only in passing, the obstacles that occur in the life of the patient with angina, and what they define as relevant to his life-world – elements of life which usually pass unnoticed in the life of a healthy person. Or consider the quadriplegic and his truncated embodiment in the world. Or even our commonplace experience of weakness and strain attendant to a case of the flu. Illness can call the self into question in fundamental fashion which outlines the contextual character of our existence.

VIII

The self is thus fundamentally a reflexive presence essentially integral with the concrete contextual situation, but a reflexivity whose presence, to become manifest, must be "called forth" and is thus implicit, dormant, hidden, until being "awakened" as a presence-to-itself in an effort to settle the unsettled.

How is this complex even possible? Or, what sense does it make to say that self is a latent presence called forth as reflexive effort? Without claiming anything definitive, I believe that two considerations will help.

A. Consider, first, several features of "contexts," i.e., of "part/whole" relations.[6] A relatively simple example of a visual "whole" will serve – e.g., a set of three points arranged to form a perceptually obvious triangle. On the basis of what Gurwitsch and Husserl have shown, it is evidently quite inadequate to speak, say, of the triangle either as the "sum" of the parts, or as "more than" its parts, or as "reducible" to its parts. Each of the points ("parts") of the triangle has its place, thanks to its being multiply related to the other points, which in turn are placed in reference to the others as well. Thus, each part is necessarily referred to itself by way of its relations to the other parts and their relations to the first – forming, thus, a system or context of mutual references. Each "part" or "constituent," in Gurwitsch's term, has its specific *functional significance*, and this significance is nothing but that complex set of mutual, self-referential interrelationships, the totality of which is what it means to speak of a "whole," or a *contexture*. "Parts" are parts of wholes strictly because they are multiple, self-referentially interrelated to one another, and to the whole thereby formed; "wholes" are wholes precisely as the complex totality of these functional significances.

The "relation" of parts and wholes is not only *complex*, but is essentially *self-referential* in character. More complex contextures would reveal still other distinctions, such as that between "self-sufficient parts" and "non-self-sufficient pieces" which Husserl analyzed. Moreover, not only are contextures more or less complex in that sense, but also there are different *kinds* of contextures, which reveal still other distinctions, such as between "parts" and "members," relatively "stable" and "unstable" wholes, or wholes which endure through moments of time and those which are themselves temporal, and still other distinctions. Such distinctions, it might also be noted, are directly relevant to a host of clinical and biomedical issues. I merely mention here Dr. Burns' point concerning "dis-ease" and "ease," and the sense in which one can sensibly speak of "parts" of an organism or mental processes as "dis-eased."[7] Or, again, the essays by Toulmin[8] and Wartofsky[9] could be fruitfully analyzed to show how such notions as "complexity," "level," "function," "system," or "whole" can be understood in the light of the phenomenon of contexture. The logic of context seems to speak directly to the conceptual clarification and explication of such notions.

A kind of reciprocity and self-referentiality is inherent to contextures as such, which either become enriched or diminished if one or another part, piece or member is added or taken away, which accordingly results in a quite different contexture – thanks to the enhancement or decrease in the set of mutual, self-referential interrelationships.

B. Returning now to the example of the traffic jam, and keeping in mind what was already noted there, a second consideration will be helpful. It is clear that there is much going on more or less at the same time: I see the vehicles, the flashing traffic light indicating it has malfunctioned, the brake lights of cars ahead of me, the traffic behind, the pavement, etc.; I hear horns honking, the grinding of gears, the sounds of tires on pavement, the wind whistling through my partially open window, the radio music, etc.; I feel the wheel in my hands, the pedals under my feet, the fabric of my clothes, the cushions of the car seat, the heat of the sun on my face, etc.; I smell the typical sorts of smells, taste the pipe tobacco, etc.; I am thinking of what to do, wishing the light worked, remembering an appointment and the thousand tiny things which bubble up, fade in and out of memory and anticipation; and on and on. Clearly, I am not equally busied with the objects of all these perceptions, memories, plans, wishes, desires. Some focally occupy me, others are more or less marginal, still others are not at all focal right now, and so on. There are, we could say, many awarenesses of many things going on, only a very few of which am I alertly focused on, or whose objects occupy me.

Let me propose that we shorten the focus of our attention, pick up merely on one of these awarenesses and its object, keeping in mind that this isolation of, say, seeing the flashing red traffic light, has its place strictly within the total context. It, like the rest, is distinguishable but inseparable. One can notice that this visual awareness of the light is itself quite complex. Merely to suggest this, the light I see is not merely red, but is experienced at the time by me as a traffic light, one which has malfunctioned, is an interruption to my prevailing plans, is seen by others, was made by determinable processes for certain determinable purposes – and all this, and more, is essentially "what I am visually aware of."

This seeing is complex in still another way: the awareness in a sense "goes on" whether or not I actively attend to what is presented through it; but now, in this example, I do attend (with dismay, anger, irritation, anxiety, etc.) to what is visually seen, hence the awareness, if you will, is "activated" by me. Or, as Husserl would say, I "live in it," actively occupied with the objects presented through it. At the same time, within the visual seeing of the light, there is a kind of implicit awareness of my eyes and more generally of my body-movements, attitudes, postures, etc. I focus my eyes, turn my head to center the light in my visual field, shade my eyes with my hand (activating certain kinaesthetic patterns of movement), shift my torso so I can be more comfortable, etc. Though not actively attentive to all these, at least with equal alertness, there are these awarenesses as well.

Leaping over still more complexities inherent even to this relatively simple case, I can note (shortening my focal attention further) that the awareness has a duration; it lasts through a certain span of time. Indeed, it is a temporal flow textured with multiple emotive, volitional, valuational, corporeal, ideational and perceptual shadings and emphases. I may shorten my attentional focus still further and consider now, say, merely a short span of this highly textured, temporal and visual awareness. To make a manageable theme here, I may also disregard everything but the specific stratum, "squinting the eyes while seeing the light." Thus specifying my focal attention, then, there is nevertheless an intriguing complexity: there is the actual *visual apprehending* of something which is enacted by specific *corporeal schemata* (focusing eyes, squinting, turning head, etc.), and the something seen which is functionally co-related with the activating corporeal schemata (what is seen, is seen thanks to those corporeal patterns embodying the strivings and seeings).

On the one hand, then, are specific mental or psychic processes having their own specific temporal flow; second, there are specific corporeal enactments of these processes; third, there is what is thereby apprehended visually. *Just here, though, arises the critical question:* the awareness is one in which "I" actively "live," busied with the "objects" thereof. In what way do "I" appear in this complex? How does this complex become, as we say, "my" experience of the traffic light? Some scholars have it that the "self" or "ego" is a sort of subjective pole for this complex, issuing from it like rays from a light-source; others that it is an underlying "substantial entity" which "has" these experiences; others that it is a sort of abiding unity which unifies what otherwise would be unrelated bits and pieces; still others that "self" is either a mere invention seductively posited thanks to language, or is an object in the world for consciousness quite as much as any other object.

It could be shown (and has, I think, by Gurwitsch for one)[10] that it is completely unnecessary to posit some underlying "pole," principle of "intrapsychic unity," or principle of unification more generally, much less an underlying "substance." The unity and continuity of such an experience is perfectly understandable in its own terms – the nature of temporality, embodiment and the theme/field character of what is experienced are quite adequate for this account. I agree with the thrust of this, but find it seriously inadequate to the issue here: whether or not one posits such entities or principles, *how does it happen not only that such an experience seems clearly to be "mine," but more directly that "I" come to "live-in" the embodied seeing and become aware of its specific object or objects? What is this "I," or "self"?* In different terms, whose "life" is it which medicine is so prominently concerned to "save"?

IX

Consider the following. It is evident, I think, that whatever appears as the object of an awareness (e.g., the light as seen) is strictly *correlated* to it: the seeing is a seeing-of-a-red-light, and the light is seen-by-the-seeing. Husserl described this "correlational a priori"[11] with his concept of *"Intentionalität."* Every awareness (noesis) is a determinable orientation towards something (noema). It is also evident, I think, that the objective correlate (noema) is functionally dependent for its specific appearance as such on particular corporeal schemata, and that the subjective awareness (noesis) is intentively oriented towards the noematic correlate by means of or through these same corporeal schemata. Thus: the light is seen, thanks in part to there being the process of seeing, but more adequately expressed, thanks to the various bodily attitudes and movements which at once enact the seeing and bring the light to focal center.

These corporeal schemata could, and at some point must, themselves be more carefully studied, especially in view of their critical orientational role in experience. For the present, however, this project must be ignored. What is prominent already is sufficient: namely, that *noesis, noema and corporeal schemata constitute a genuine contexture of their own.* Each is a veritable constituent bearing in itself the full significances already delineated, but with several further and crucial differences. The contexture mentioned earlier – the three points forming a triangle – *is strictly an object,* a noematic correlate, and the self-referentiality and reciprocity noted pertain to the sense in which it is an integral whole. Here, however, we are confronted with a far richer contexture, and this richness can be specified. First, the "what is seen" is itself a contexture like that analyzed earlier, but is here only one constituent of a contexture. Second, it turns out that the corporeal schemata pertain to the embodying organism, and thus are themselves constituents of a complex contexture (with its own appertaining "parts," "members," and the like), as, third, so are the noetic awarenesses embodied through these schemata. Such awarenesses, fourth, constitute a temporal flow *(Erlebnisstrom).*[12] Beyond these dimensions – each of which would have to be studied as contextures – is another. The sense in which noema, noesis, and corporeal schemata are specifically enriched and thus reciprocally constituted *as* constituents of an integral totality, is that *each is functionally significant vis-à-vis the others*. Thus, one way of expressing this complex contexturing is to say, as already suggested, that the awareness of the light is at the same time an awareness of *itself as aware of the light*, and *an awareness of the embodying corporeal schemata* as such.

Sartre seems to have seen this complexion, noting that every consciousness *of* something is at the same time a consciousness (of) itself as that consciousness which it is.[13] He places brackets around the second "of" to indicate that this awareness (of) the awareness is, as he says, a non-reflective cogito, which is the necessary and sufficient condition for there to be the awareness in the first case. To call it non-reflective, however, just as clearly obscures the very point at issue: that this (of) indicates the reflexivity constitutive of any awareness whatever. In his earlier *The Transcendence of the Ego*,[14] which purports to empty consciousness of any "ego" and place it in the world as an object like any other, but ends up calling the ego "fugitive" and "haunting" – a magical phenomenon even – he seems to me to have been much closer to the truth. For "self" is precisely *no-thing*, no substance within or behind consciousness, but neither is it a wordly object like grapes and stones. It is indeed "fugitive" and "haunting": self, I submit, is neither noema, noesis, nor corporeal schemata, but what I might now call the *reflexivity* of the contexture itself.

X

Even this is still not adequate, however, for it does not yet capture the uniqueness of this complex set of constituents. Two points are essential. First, the corporeal schemata at once embody the specific noeses (seeing, squinting, focusing, etc.) and functionally determine the appearance or presence of the noema: they at once "flesh out" the noesis, and allow actual presence to the noema. More generally expressed, the animate organism actualizes awarenesses (noeses) *by orienting them towards objects* (noemata), and organizes the experienced field of objects with respect to the animate organism. The latter is, as Husserl points out, the fundamental *orientational point* with reference to which objects (noemata) are displayed, arranged or organized as "here," "over there," "left," "behind," and the like. It is the actional center, the locus of experience, and the orientational reference for correlated objects. The embodying animate organism, then, itself displays a singular kind of reflexivity, expressed in this phenomenon of double functioning – a reflexivity which persuades thinkers like Merleau-Ponty that the body itself is "a subject," a "self."

Second, the noetic component of the stream is intentive, or *is positioned with respect to*, the noematic component. Experience, the concatenation of multiple awarenesses, however, is a *temporal* flow, highly articulated, as Husserl shows,[15] into complex and interconnected phases each of which is simultaneously protentive (anticipational), retentive, and impressional. If one considers any one phase, it becomes immediately clear that, as a phase within a temporal flow of phases, it is constituted as a "now" (impressional) by virtue of its being retentive to past phases of itself as having been phases which then (as impressional) protended the present phase (among others) as going to occur and (when occurring) as going to be retentive in this specific sense (retaining the past phase as a phase protending the present as going to be retentive of the past phase as ...). At the same time, the "now" phase is protentive of phases not yet occurring but which, when they occur, will be retentive of the present now-phase as having been protentive in this specific sense. In more general terms, because of this complex structure of inner-time consciousness, I can now remember that yesterday I expected to remember yesterday; and, I now expect that tomorrow I shall remember that today I reminded myself to do something tomorrow.

Each phase, thus, is retended or protended *in all its complexity*, i.e., each phase preserves its impressional/retentive/protentive structure. There occur, thus, *syntheses of transition* among the noetic temporal phases, but these transitions from phase to phase are, first, *continuous*, and second, as Husserl emphasizes, are fundamentally *syntheses of identification*. Each phase of the temporal flow of noeses is what it is thanks to the multiple self-references inherent to these retentive and protentive characters: "now" is constituted as "now" by virtue of the multiple retentive and protentive references. The noetic flow, thus, is as well a complex totality of reflexively interconnected references.

This noetic reflexivity, however, beyond its own peculiar temporal character, is intentively positioned with respect to the noematic correlate (to "what appears" or is somehow experienced) strictly by way of the embodying organism. Hence, not only are there awarenesses of objects (seeing the traffic light), but these awarenesses are corporeally positioned, oriented, or situated. As such, the *noetic is always complexly intentive*:

towards the noematic correlate, but strictly by means of embodying schemata. Furthermore, the specific sort of reflexivity of the latter is thus similarly complex or enriched – in itself, with respect to the noematic correlate (to which there is the awareness), and as corporeally positioned and oriented.

What thus becomes prominent in these considerations is that the reflexivity displayed by the embodying organism's schemata (attitudes, movements, sensory fields, etc.), and the complex reflexivity inherent to the noetic phases of awareness, are essentially positioned, or *situated*. The reflexivity constituting this temporally ongoing, corporeally embodied and intentively oriented flow of experience *has a locus, a "where-at"*; it is a *habitus*. *"Self" is fundamentally a situated or positioned reflexivity oriented towards the world which is itself displayed, arranged, or organized strictly in reference to this reflexively oriented habitus*. The self is literally no-thing; nor is it identical with any of the delineated constituents. Neither noesis, organism, nor noema, *self is precisely the peculiarly complex reflexivity itself: that by virtue of which any reflexive referencing* ("itself," "myself," etc.) *is at all possible*. In that sense, self turns out to be the *eidos* of human life.

XI

A final word is necessary, lest the many threads of this too loosely woven tapestry become unraveled (if they are not already). The examples and deliberations developed earlier made prominent the phenomena of alertness, effort and possibilizing – prominent as such in the context of upset, disturbance, impasse, and the like. Without attempting here to carry the analysis to the fundamental levels it doubtless should be – time is at a premium, and insight fails – I want to suggest some ways of understanding the emergence of self (situated reflexivity) as an alerted efforting expressed in contexts calling for possibilizing. I want also here again to stress the important ways in which illness can adumbrate the nature of the self. If the self is a peculiarly complex reflexivity, a reflexive referencing called into effort by the hindrances of life, then illness evokes, defines, delimits, and truncates the self. Illness upsets the "taken for granted" routines of the everyday life-world and places the ill person in a world of illness, defined by the character of his disease. Alvan Feinstein's suggestion [16] that we study the character of illness should also be combined with a sensitive attention to the phenomenology of the various illnesses, an attentive description of the significance of illness, *for* the self that lives *in* and is defined *by* an illness.

Both Piaget and Husserl, and others, have emphasized that the fundamental character of life (biological as well as psychical) is the tendency or impulse to maintain or establish itself as a harmonious equilibrium. In Piaget's terms,[17] for instance, the movements of even the most rudimentary reflexes (blinking, sucking, grasping) tend to function repetitively, so to speak, just for the sake of functioning. This inherent tendency is to assimilate the environs to themselves. It is only when reflex movements come across something which cannot be assimilated, or can only partially, that there occurs an accommodation of the movements to that something. With that, furthermore, there occurs an internal organization of movements, a crystallization and consequent differentiation of movements into schemata having, one could say, the germinal sense of appropriateness: only some affairs can be grasped, for instance, and to bring non-graspable things into the reach of the organism's ongoing life, other schemata must come into play. To assimilate, the organism must accommodate itself to its environs, thereby organizing it; this brings into play the coordinate crystallization of schemata – all in the service of maintaining equilibrium. Husserl's account,[18] it could be shown, is remarkably similar, even though there are substantial differences, to be sure.

I cannot here, for lack of time, engage Piaget's account critically. I nevertheless propose that it is suggestive to adopt and adapt its basic terms to the present theme.

It seems sensible to speak of even reflexes as embodying noetic intendings (strivings, wantings, etc.), however rudimentary, diffused, or global they doubtless are. To that extent, the complex *habitus* of interlaced and reflexive connections noted is already operatively functioning, but its basic impulse is to maintain or preserve harmonious equilibrium or balance. Differentiation and individuation occur on the basis of upsets to or in the equilibrium; or, negation occurs only in the context of affirmation (positioned equilibrium), and leads to differentiation and individuation. One might think here of Adolf Meyer's life charts, in which physical and mental disturbances define the development of an individual.[19] So long as no upset occurs, there is, if you will, relative homogeneity, a relative absence of differentiation in the field of awareness, the corporeal functioning and the noetic flow. A disturbance in any of the constituent components, however – whether something appears in the field discordant with expectations, a bodily process undergoes change and may thus fail to embody, or the noetic flow alters and finds no adequate enactment in bodily schemata – signifies an upset in the prevailing equilibrium, and thus prompts or evokes an accommodation (or what I have called differentiation and individuation).[20] So to speak, there arises a sort of "having to do something" as a consequence of the disturbance, in order then to re-establish, re-arrange, or re-situate.

More particularly, an upset signifies an alteration or relative break in the hitherto prevailing relationships among the constituents, and thus *signifies a modification of some order in the constitutive reflexive orientations among them.* Precisely to the extent that this ongoing "life's" tendency is to maintain equilibrium – which means, *to maintain itself as such* – this modification in the contexture provokes an "alertness," or in Husserl's terms there is a "solicitation" (by the unsettled and unsettling context) for "ego-advertence." To re-establish the contextual balance or equilibrium, "effort" is elicited. But, for the re-orientation to come about, this effortful alertness must emerge as an *advertence to* "what has happened," to what is unsettled and unsettling, and this advertence has the core sense of at-tending: noticing and attempting to tend to. That is, there is not simply a sort of passive noticing, but a quite active tending and attending precisely because, granted the rudimentary tendency to maintain contextual balance, *it is exigent that "something be done"* – how exigent will depend upon how deeply the disturbance disturbs, how upsetting is the upset, how seriously in short the prevailing equilibrium is interrupted or broken. I suspect, but only that, that central concepts like "disease" and "health" can be illuminated in this light.

This notion of the self reveals medicine as a unique avenue for understanding the self, for seeing the ways in which it is both nurtured and destroyed by adversity. Medicine offers variations upon the achievement of the self by offering contexts that grade from the everyday life-world of the normal adult into the contorted and reshaped contexts of the ill. Some of these contexts we have all visited – the delirium of fever, the pain of a headache, the transient paralysis of a limb that has gone asleep.

Others offer special vantage points upon the limits of the life of the self
– consider, for example, the schizophrenic whose life is sundered by
failing associations, isolated by autism, rendered distorted by ambivalence and inappropriate affect. In short, the appreciation of the contextual
character of the self indicates how the self will take on different proportions in illness and health, while indicating at the same time the need to
take these considerations to a phenomenology of illnesses.

To be more general, from alertness and effort, to advertence as exigent
attending, the focus seems to be one which *seeks*, specifically *seeks to
reckon with and account for* what has disturbed. Such reckoning and accounting, furthermore, seem basically the exigent effort – willful striving
– to re-orient or re-organize by means of incorporation into prevailing
schemata (where possible), or by modifying these. Failing that, the search
assumes the thrust of seeking other possible ways of re-establishing the
contextual balance. Hence, the alertness and effort have the significance
of "de-actualizing," "taking account of," or "thinking for the possibly
otherwise," i.e., *possibilizing* – the critical move from what is actual to
the "possibly otherwise." With this, finally, there begins to occur the kind
of differentiation which I think is the sense of Piaget's notion of crystallization into schemata – differentiation within the noetic flow, the corporeal schemata, and within the noematic correlate, and because of this,
differentiations of the modalities of reflexivity obtaining among these
components. Hence, as Piaget goes on to note, "self" and "object" become progressively differentiated from one another – terms which, however, it should now be clear, cover over as much as they illumine.

XII

Such disturbances (impasses, crises, and the like) set in motion this complex multilinear and reflexive search focused on accounting for and thus
re-establishing. In this sense, we have to do with contextures. That is,
unless there were the specific kinds of mutual and reflexive interrelationships – if these were merely relations *partes extra partes* – no amount
of alteration or modification would prompt such alertness, effort and
the like. The phenomenon in question does display essentially contextual significances, however, and this suggests that such changes are
always other than mere alterations or modifications: they are genuine

disturbances, hence solicit alertness, effort, attending, and the rest.

Clearly, much has been left out of this – strata and dimensions of self, person, human life – which is crucial. I think, though, that these suggestions go a long way toward understanding the genealogy of self, its fundamental situational locus, but also its terribly baffling and fugitive character.

To conclude in somewhat different terms, *self as situated reflexivity is a contextual phenomenon; it is precisely that by virtue of which the complex of reflexive interconnexions of temporally flowing noeses and embodying corporeal schemata become oriented, positioned as a habitus in the midst of environing things, objects, persons, ultimately the world.* Its emergence has the sense of an "awakening," called forth and having the forms of alertness, effort, exigent attending, reckoning, accounting, possibilizing and mutual differentiation. For self to emerge into explicit wakefulness, these considerations suggest, requires the occurrence of disturbances, upsets, crises. Thus, more generally, the phenomenon of self is in essence a problematic one, in the specific sense that self is in essence always a *problēmā* to itself. The problem of self is that the self is a foundational problem to itself. Its emergence and prospectively enriched modes of embodiment thus are strictly dependent upon the occurrence of the unexpected, with respect to which the striving to maintain emerges by possibilizing efforts resulting in increasing differentiations of noetic and corporeal schemata, and noematic strata. The genealogy of self thus signifies the emergence of heterogeneity, complexity, enriched reflexivity, from relative homogeneity.

This incredible complexity to which the notions of reflexivity and contexture are directed concerns only one facet. The ways in which self relates to other selves have not even been hinted at here, much less the ways in which this situated reflexivity becomes articulated in moments of personal choice, decision, etc. I have tried merely to grapple with the foundational aspects of this complexion, only archaeologically outlining some of the prominences on this terrain. Further expeditions are in order, but possible only with more refined tools, greater patience, and many more subtle skills.

I thus conclude with more caveats than solutions for particular uses of self in medical and psychiatric discourse. The self's nature is a delicate fabric. But even to learn this is to learn at least where to begin, to be recalled to a critical awareness of the fragile enterprise which is at our core as persons and thus at the core of our concern about persons and those persons who are patients. Even this is a beginning. Medicine can offer to phenomenology a field of experience that gives illuminating variations upon the significance and existence of the self. What I have done here is but to lead up to that point and invite others to follow into the issues themselves.

Southern Methodist University,
Dallas, Texas

NOTES

[1] William Golding, *The Spire* (New York: Pocket Books, 1966), p. 4.

[2] T. S. Eliot, "Burnt Norton," in *Four Quartets* (New York: Harcourt Brace Jovanovich, 1943), pp. 121–2.

[3] *Vide* R. M. Zaner, *The Problem of Embodiment: Some Contributions to a Phenomenology of the Body, Phaenomenologica* 17 (The Hague: Martinus Nijhoff, 1964).

[4] William Heberden, *Commentaries on the History and Cure of Diseases* (Boston: Wells and Lilly, 1818), p. 293.

[5] H. Tristram Engelhardt, Jr., *Mind-Body: A Categorial Relation* (The Hague: Martinus Nijhoff, 1973), p. 2.

[6] *Vide* the pioneering analysis of part/whole relations by Edmund Husserl, *Logical Investigations*, trans. by J. N. Findlay from 2nd ed., 2 vols. (New York: Humanities Press, 1970), and that of contexts by Aron Gurwitsch, *Field of Consciousness* (Pittsburgh: Duquesne University Press, 1964).

[7] Chester R. Burns, "Diseases Versus Healths: Some Legacies in the Philosophies of Modern Medical Science," in this volume, p. 29.

[8] Stephen Toulmin, "Concepts of Function and Mechanism in Medicine and Medical Science," in this volume, p. 51.

[9] Marx Wartofsky, "Organs, Organisms and Disease: Human Ontology and Medical Practice," in this volume, p. 67.

[10] Aron Gurwitsch, "A Non-egological Conception of Consciousness," in his *Studies in Phenomenology and Psychology* (Evanston: Northwestern University Press, 1966), pp. 287–300.

[11] Edmund Husserl, *The Crisis of European Sciences and Transcendental Phenomenology*, trans. by David Carr (Evanston: Northwestern University Press, 1970), §46.

[12] Edmund Husserl, *The Phenomenology of Internal Time-Consciousness*, ed. by Martin Heidegger, trans. by James S. Churchill (Bloomington: Indiana University Press, 1964).

[13] Jean-Paul Sartre, *Being and Nothingness: An Essay on Phenomenological Ontology*, trans. by Hazel E. Barnes (New York: Philosophical Library, 1956).

[14] Jean-Paul Sartre, *The Transcendence of the Ego*, trans. by Forrest Williams and Robert Kirkpatrick (New York: The Noonday Press, Farrar, Straus and Giroux, 1957).

[15] Husserl, *The Phenomenology of Internal Time-Consciousness*.

[16] Alvan Feinstein, *Clinical Judgment* (Baltimore: Williams and Wilkins Company, 1967), p. 126ff.

[17] Jean Piaget, *The Origins of Intelligence in Children*, trans. by Margaret Cook (New York: International Universities Press, 1952).

[18] Edmund Husserl, *Experience and Judgment*, rev. and ed. by Ludwig Landgrebe, trans. by James S. Churchill and Karl Ameriks (Evanston: Northwestern University Press, 1973).

[19] *The Collected Papers of Adolf Meyer*, ed. by Eunice E. Winters (Baltimore: The Johns Hopkins Press, 1952), vol. 4, *Mental Hygiene*, "Mental and Moral Health in a Constructive School Program," esp. pp. 352–53. Meyer's proposal to study the total, integrated, and unified individual, it seems, is one which depends precisely on understanding human life as a *contexture*, and his idea of life charts clearly reveals the beginning of an appreciation of the logic of wholes and parts, i.e., of contexts.

[20] Bruno Bettelheim, *The Empty Fortress* (New York: The Free Press, 1967), esp. pp. 20–34. Bettelheim's study of autism shows unmistakably the significance of alertness, effort, and reflexivity for the emergence of self – or its attenuation in the case of autism.

ANDRÉ SCHUWER

COMMENTS ON "CONTEXT AND REFLEXIVITY"

Professor Zaner's paper is appropriately subtitled a "Genealogy of Self," for it represents an attempt to exhibit the emergence of the subject or self. He does this in language which does not presuppose – in the first part of the paper – direct acquaintance with any of the numerous philosophical traditions in which the "self" has in modern times been treated. To comprehend the phenomenon of "self" is a pressing task for us. Such a comprehension will enable us "to account for the specifically human" and "every discipline focusing on 'the human' and 'the human world' – social and behavioral science, history, theology, *as well as the biomedical sciences, medical practice*... finally depends for its sense and justifiability on the elucidation of the phenomenon of 'self'" (pp. 155–156, emphasis added). Zaner clearly intends to speak to the issue of this Symposium in pointing out, in the text we just quoted, the role philosophy should play in the biomedical sciences. He is well aware that philosophy does not enter into this role out of a position of strength.

There is a wide range of terms used to indicate the reality which the expression "self" signifies. These terms denote conceptions of "self" which are very different, often opposing and contradicting each other. This linguistic and conceptual confusion makes it necessary to return to "the things themselves." Our thinking has "to shape itself in accord with *what self is*" and must be "accurate, fair and adequate to what we want to think about and thereby account for" (p. 156, emphasis added). To gain access to the phenomenon of "self," Zaner endeavors to "let it stand out for itself and become evident on its own" (p. 156). He engages in a descriptive explication of a concrete situation – being caught in a traffic jam, while underway to work, and trying to get out of it – which he analyzes in order to find what we mean when we use the term "self." This analysis occupies close to half the paper and Zaner is here at his very best. This analysis recalls his well-known book, *The Way of Phenomenology*, where he introduces phenomenology "by actively engaging in it."[1] Zaner invites us to learn how it is meaningful to speak of the

H. T. Engelhardt, Jr. and S. F. Spicker (eds.), Evaluation and Explanation in the Biomedical Sciences, 175–180.
All Rights Reserved. Copyright © 1975 by D. Reidel Publishing Company, Dordrecht-Holland.

"self" by "actively engaging" in a lived context in which such a reality, which we signify by the term "self," is called forth. This approach makes it difficult to discuss the paper in terms of its conclusions, since these conclusions cannot be abstracted from the way (method) through which he arrived at them. We will therefore first give a summary of the steps of the analysis and then point out some of the questions which, according to our understanding, issue from it.

(1) In the analysis of the concrete situation of the traffic-jam, we can distinguish five stages. The first stage of this analysis concludes to the facticity of "referrals." Also included here is the fact that what I do at any moment is part of an "integral context" (p. 157). The relation of one's involvement to the "integral context" leads Zaner to conclude, regarding human being in general, that "Human life is a network of mutually interlocking relationships" (p. 159). In the concrete situation of the car-driver who is caught in a traffic-jam it must be underlined that the incident *interrupts* 'the usual taken-for-granted course of affairs" (p. 156). The car driver must "think" and must look for a way out. This means a "de-actualizing" of the situation in order to actualize one or another alternative which the situation allows for. Zaner then "suspects" that "the course of any human life always involves just such situations" (p. 159). The car driver is capable of being interrupted. And this signifies that he is capable of being alerted and of acting out efforts to pursue his course. It is pointed out in this context that "... such alertness and effort ... must of necessity also be reflexive" (p. 159). A creature which is "locked to the actual" cannot be called a "self." Concerning the "self," Zaner then maintains: 'Self' makes its appearance as a *reflexive alertness and effort* ... within concrete contexts of concern and action" (p. 159). This "self," moreover, is one which involves and *always* involves bodily enactment. Zaner thus comes to the fifth stage of his analysis when he presents a fuller definition of "self": "The self is thus ... a reflexive presence essentially integral with the concrete contextual situation, but a reflexivity whose presence, to become manifest, must be 'called forth' and is thus implicit, dormant, hidden, until being 'awakened' as a *presence-to-itself* in an effort to settle the unsettled" (p. 161).

(2) It must be noted here that, at this point of the analysis, he has worked his way up from a concrete example of human situation and human action to a rather full-fledged conception of that "self" which

began as so problematic at the outset of his paper. At the same time, his language becomes more philosophic; it takes on the distinctively philosophic character which he earlier avoided. Now "a reflexive presence" and a "presence-to-itself" have emerged as issues. With Section 8, Zaner interrupts the discussion of his example of the traffic-jam which he then illustrates in medical terms with reference to a person with angina. He then considers "several features of 'contexts,' i.e., of 'part/whole' relations" (p. 161). He gives the example of a visual "whole," that is, "a set of three points arranged to form a perceptually obvious triangle" (p. 161). He finds that not only is each point related to the other points; beyond this, each is "self-related," by virtue of that relation. Only by virtue of this compound of relations is the whole a whole. Zaner writes: "The 'relation' of parts and wholes is not only *complex*, but is essentially *self-referential* in character" (p. 162). The point to the discussion here is that "a kind of reciprocity and self-referentiality is inherent to contextures as such" (p. 162). That is to say that the same kind of reciprocity and self-referentiality applies to *all* cases of "wholes" ("contextures"). Zaner then applies this to his traffic-jam example. He finds it necessary to concentrate on just the traffic light blinking, in an effort to see what "goes on" here. "I ... attend to what is visually seen, hence the awareness ... is 'activated' by me. Or, as Husserl would say, I 'live in it' .... At the same time, within the visual seeing of the light, there is [an] implicit awareness of my eyes and more generally of my body-movements, attitudes, postures, etc. ... [Moreover] I can note ... that the awareness has a duration; it lasts through a certain span of time. Indeed, it is a temporal flow with multiple emotive, volitional, corporeal, ideational and perceptual shadings and emphases ...." (p. 163). Zaner draws three distinctions from this: "There is (a) the actual *visual apprehending* of something which (b) is enacted by specific *corporeal schemata* (focusing eyes, squinting, turning head, etc.) and (c) the something seen which is functionally co-related with the activating corporeal schemata" (p. 164).

(3) At this point, the strict analysis ceases and he asks the question he set at the beginning of the paper: "In what way do 'I' appear in this complex?" (p. 164). He suggests that it is not necessary to posit "some underlying 'pole' ... or principle of unification .... The unity and continuity of such an experience is perfectly understandable in its own terms ...." (p. 164). However, he judges that such an answer is inadequate, since it

avoids confrontation with the question: "How does it happen... that 'I'
come to 'live in' the embodied seeing and become aware of its... object...?" It is also at this point that Zaner's debt to a Husserlian manner
of accounting for what occurs here, and eventually for the "self," becomes
obvious, as he redescribes the terms of the analysis in Husserlian expressions, noesis and noema. He writes, "... *noesis, noema and corporeal
schemata constitute a genuine contexture of their own*" (p. 165). The role
of these "corporeal schemata" seems rather greater than that of the other
two components here, the corporeal schemata being the "bodily attitudes"
of the perceiver. What Zaner does is to locate on the higher, more complex
level of interrelations between noesis, noema and corporeal schemata
the same interdependence he found earlier within the "parts" of the triangle. After considering an object from the Sartrian point of view – Sartre
intending "to empty consciousness of any ego" (p. 166), Zaner finds that
the "self" is "neither noema, noesis, nor corporeal schemata, but what I
might now call the *reflexivity* of the contexture itself" (p. 166).

(4) But this finding is not yet sufficient: "... it does not yet capture the
uniqueness of this complex set of constituents" (p. 166). The improvement on his preliminary definition of "self" consists in a wholly Husserlian discussion of the three components of the contexture in its reflexivity.
We will pass over this very interesting discussion to mention his conclusions: "*'Self' is fundamentally a situated or positioned reflexivity oriented towards the world which is itself displayed, arranged, or organized
strictly in reference to this reflexively oriented habitus*" (p. 168). By "*habitus*"
Zaner has indicated he means the point or *locus* from which reflexivity occurs. In brief, *"Self is precisely the peculiarly complex reflexivity
itself: that by virtue of which any reflexive referencing ... is at all possible"*
(p. 168).

(5) This conclusion is still not complete as far as Zaner is concerned;
he follows it with four pages of observations on what he calls "understanding the emergence of self" as "an alerted efforting expressed in
contexts calling for possibilizing" (p. 168). These discussions attempt to
show that "self" is involved even in situations presumed solely "reflexive." Zaner alludes here to researches of Jean Piaget and relates his
further clarifications of the concept of "self" to Piaget's (and Husserl's
as well) emphasis "that the fundamental character of life (biological as
well as psychical) is the tendency or impulse to maintain or establish

itself as a harmonious equilibrium" (p. 169). And we should recall here that Zaner had already, in the early stage of his analysis, made the observation that interruptions, impasses and disturbances call forth "self" out of its dormant state.

At the beginning of his paper, Zaner says that the project he has set himself is "mammoth and unmanageable" (p. 153). At the end of his paper he speaks of the "incredible complexity to which the notions of reflexivity and contexture are directed" (p. 172). Moreover, he is well aware that he did not carry his analysis of reflexivity and contexture "to the fundamental levels it doubtless should be" (p. 168) on which "self relates to other selves" (p. 172) or on which "situated reflexivity becomes articulated in moments of personal choice, decision, etc." (p. 172). With this in mind we will now point out some of the questions which, according to our understanding, issue from our careful reading of Zaner's paper. We hope that we did not misunderstand him.

1. On page 165 Zaner writes: *"Noesis, noema and corporeal schemata constitute a genuine contexture of their own."* This contexture must be conceived according to the model of the contexture of the three points, a visual whole, the perceptually obvious triangle. We might ask: What, in the case of the traffic-jam as contexture, is supposed to be analogous to the triangle? Zaner claims that the analogue between the triangle and the contexture of the traffic-jam seems clear. But we have a problem in understanding just what exactly is alleged to equate to the triangle (a case drawn, it should be noted, not from perception but from conception, although Zaner might deny this). It should furthermore be asked whether Zaner's analysis of the part-whole relations in the visual example of the three points of a triangle is correct. Reciprocity and self-referentiality are characteristic for this contexture. It is essential to them. He says that not only is each point related to the other point, to its left and to its right, but beyond this each is "self-related." We would like to ask here in what sense "self-related" is meant? What kind of understanding of "self" is operative here? Is a point a "self?" If the self-referentiality, which is characteristic of the contexture of the traffic-jam, is like the self-relatedness of any apex of a triangle, and if the self-relatedness of this apex is rather unclear as to its precise meaning, what is then the meaning of the self-referentiality of the contexture of the traffic-jam? Furthermore, the example cannot be geometrically satisfactory, we think. Is not all he has

given here, *not* the parts of the *whole* triangle but *only* the apices of the triangle? So we have not considered a whole being.

2. At the end of his paper, Zaner seems to give his assent to a thesis of Piaget that "the fundamental character of life (biological and psychical) is the tendency or impulse to maintain or establish itself as a harmonious equilibrium" (p. 169). In the case of the traffic-jam we can understand that life rebels against a traffic-jam which disturbs the harmonious equilibrium of the car driver underway to his work. But is this apparently "eidetic" affirmation about life founded in the description? Are there not crises which "call forth" the "self" and which call it forth to *surpass* itself? Are there not progressive forces of life which battle against the concern to maintain a harmonious equilibrium, a concern which ever so often shows itself to be an obstacle to growth of "self?" Are there not challenging changes which make us tap resources hitherto hidden and unknown to us?

3. Zaner speaks of the "inner *logos* of self's emergence" (p. 153). His project of a "genealogy of self" purports to seek this inner logos. We would certainly like to hear more about this inner logos. Does Zaner allude here to some kind of contextual logic of our bodily perceptual anchorage in the world in which "self" is hidden and called forth? What kind of "logic" rules the "oddly fugitive reflexive presence of "self?"

Professor Zaner speaks furthermore of the "self" as "the *eidos* of human life" (p. 168). Referring to Merleau-Ponty, Zaner writes that "the body itself is ... a 'self'" (p. 167). Might we not conclude then that "self" is some kind of protean entity, some kind of multifaceted entity, which can hardly satisfy our medical colleagues who have first been told how important an elucidation of the phenomenon of "self" is for the justification of their biomedical science?

Duquesne University,
Pittsburgh, Pennsylvania

NOTE

[1] Richard M. Zaner, *The Way of Phenomenology* (New York: Pegasus, 1970), p. xii.

STUART F. SPICKER

THE LIVED-BODY AS CATALYTIC AGENT:
REACTION AT THE INTERFACE OF MEDICINE
AND PHILOSOPHY*

I. INTRODUCTION

Throughout the history of Occidental philosophy one generation after another (including philosophers as well as physicians) has had to suffer the tyranny of spiritualistic metaphysics, what Edmund Husserl, the founder of 20th century phenomenological philosophy, called "histori-cally degenerate" metaphysics.[1] I hasten to qualify this assertion by add-ing that such suffering, in my judgment, need not be alleviated by the questionable practice of euthanasia on the part of contemporary philos-ophers, although one may well make the case that negative euthanasia[2] is justified in the case of a speculative metaphysics which has produced little more than an emptily formal ontology. Yet this harsh judgment need not entail the conclusion that Metaphysics *überhaupt*, to which some physicians and philosophers have, regrettably, an aversion, be henceforth rejected and abandoned. Indeed, it is one of the aims of this essay to make plausible and palatable the claim that Metaphysics or First Philosophy is in fact intimately bound to Medicine[3] at their inter-face – the lived human body – and that this is at once identical with the initial aspirations of both philosophy and medicine, qualified by the ex-clusion of what Husserl called "all speculative excesses."[4]

As a paradigmatic illustration of spiritualist extravagance of the Oc-cidental tradition, consider a passage from The Fourth Ennead of the third century Neoplatonist, Plotinus (205–270), who in the "Eighth Tractate" introduces his analysis of "The Soul's Descent Into Body"[5] with a description of what is at once a mystical experience and a meta-physical thesis, which when taken together constitute the significance of his philosophical view of Soul and body:

Many times it has happened: lifted out of the body into myself; becoming external to all other things and self-encentered; beholding a marvellous beauty; then, more than ever, as-sured of community with the loftiest order; enacting the noblest life, acquiring identity with the divine; ... yet, there comes the moment of descent from intellection to reasoning,

H. T. Engelhardt, Jr. and S. F. Spicker (eds.), Evaluation and Explanation in the Biomedical Sciences, 181–204.
All Rights Reserved. Copyright © 1975 by D. Reidel Publishing Company, Dordrecht-Holland.

and after that sojourn in the divine, I ask myself how it happens that I can now be descending, and how did the Soul ever enter into my body, the Soul which, even within the body, is the high thing it has shown itself to be.[6]

This passage, among innumerable others, indicates that for Plotinus a person is strictly identifiable with the Soul, and the body (I hesitate to say *his* body) is eventually, at the end of its earthly life, a victim of destructive agents of many kinds, since it is a composite which cannot forever hold together, especially when "material masses are no longer presided over by the reconciling Soul."[7] But this is to enter the metaphysical dialogue without engaging the man. Plotinus' disciple, Porphyry, however, tells us something of the master himself:

Plotinus, the philosopher our contemporary, seemed ashamed of being in the body. So deeply rooted was this feeling that he could never be induced to tell of his ancestry, his parentage, or his birthplace.[8]

Such a personal stance would hardly be acceptable to empirically oriented philosophers, to say nothing of practicing health professionals committed to acquiring, through the mechanisms of medical history (including social and family history), systems review, and physical examination, a foundation for assessment of their patient's medical status. And who would have it otherwise? Initially, the physician's task, after all, is to comprehend the biological and to understand the principles of organic processes as a background against which to appreciate the body's immunological mechanisms and functions and to relate all of this and bring it into play in understanding the mechanisms and agents of disease and disease processes.[9] Already, in this last sentence, I made reference to the mechanisms and functions of "the body"[10] as if we all agreed on the meaning of this common word. Paradoxically, in spite of the emergence of the centrality and import of man's bodily being as the very subject matter (or should I say 'object matter') of medicine, one finds that we have not thought ourselves through to a more sophisticated image of the body, partially because we allow ourselves to speak of 'the body image' and other such *scheme*[11] or ghosts which, I think, we would well be rid of by adopting a method of intellectual exorcism. Troublesome metaphors for the human body (often erroneously attributed to the physician-philosopher, Julien Offray de la Mettrie and his infamous title *L'Homme Machine*)[12] include the clock, a system of levers, a heat engine, a chemical factory, a dynamo, a storage battery, a radio receiving and transmitting set, a plant that runs on sun power, a colony of cells, a textile of tissues, a writing, reading and calculating machine.

II. THE PHYSICAL BODY AND THE LIVED-BODY

It is something of a historical paradox that seventeenth-century thinkers, as Professor Toulmin has already shown,[13] "began with quite clear ideas about what matter is, what a machine or mechanism is and what was required of a mechanistic explanation," these thinkers imposing a "restrictive definition"[14] on terms like 'matter', 'machine', and 'mechanism'. So although we have to some extent abandoned the notion in philosophy and medicine in general that the soul is the proper subject matter of medicine and the medical sciences, we have yet to establish an adequate account of the human condition in terms of a strictly *bodily* account of its existence.

The philosophical tradition reflects the distinction between 'animate' and 'inanimate' existence. If inanimate, then man is nothing but a brute mechanism, insensible, and soul is needed to account for thinking. This has proved hopeless. If animate, then matter, e.g., minerals, are remnants of organic beings; this is surely false. It is to the everlasting genius of la Mettrie that he rejected the proposition that "Man is either animate or inanimate."[15] Rather, he set forth the thesis that a living man is a particular type of organization of matter [*la matière*]. This thesis has apparently found its adherents in medicine, thus giving the impression that philosophers with their metaphysics were no longer appropriate bedfellows, which was captured in such phrases as "medicine has now advanced adequate explanations of material processes as to be rid and free of any notion of soul, a pseudo-attempt to account for vitality and the animate character of living beings." Quite surprisingly, however, many philosophers on the Continent have joined biomedical scientists and physicians in this stance, but for entirely different reasons.

The consequence of rejecting 'soul talk' from discussions of the nature of biological life, especially human life, has been to assume (quite incorrectly) that such life is entirely explicable in the language of the biomedical sciences, especially molecular biology (which includes molecular genetics and cell physiology), and even immunology. This I hope to show is mistaken. An equally serious metaphysical difficulty is still with us,

notwithstanding the exclusion of *psychē* or soul from the current dialogue. So let us turn to *Soma* ....[16]

The German language reflects a distinction between two conceptions
of the human body: It has the word '*ein Körper*' to refer to the human
body as just 'a body', 'a physical body', 'the body as object', a physico-
chemical system, understandable through established laws of physics;
this is the object of investigation of the positive sciences and the biomedical sciences in particular. This is the body which is studied by anatomist and physiologist.[17] The medical student and physician is most
familiar with this sense of 'body'. J. H. Van Den Berg in his regrettably
little known *Changing Nature of Man* puts it quite vividly:

> The doctor who taps on the chest, who palpates the abdomen, who raps under the patella,
> does not touch the same body that demanded its owner's attention. He only touches the
> conditions, the instruments, of that of which the owner can dispose. Before the doctor
> begins his examination, the owner has left his body, he has abolished his incarnation; that
> is why he can offer his body as an object for examination, as "chest with caverns," as
> "abdomen with probable abscess," as "knee without normal reflexes." The help offered
> by the doctor concerns the body, this unanimated, disincarnated body, this instrument.[18]

But the German language reflects quite another conception in the word
'*ein Leib*' than this 'unowned body' which we can 'forget'.[19] This word
has quite a few French parallels: Merleau-Ponty speaks of '*le corps
propre*' [my body proper]; Gabriel Marcel discovered the phenomenon
(of which I shall say more) and speaks of 'my body as mine'. Sartre refers
to 'my body-as-lived'; others simply speak of 'my body-as-lived-by-me';
Professor Zaner, following our late teacher Dorion Cairns, translates
Husserl's use of *Leib* as 'animate organism',[20] cautioning that it should
never be translated 'living body' for the good reason that this would
sound too biological and thereby precisely miss the point intended. Yet
Husserl's notion of the living or animate organism (as the one and only
one which I rule and govern immediately in and through each of its
organs)[21] is not quite as good a description as some have claimed it to be.

The lived-body is not apprehended in its full sense when rendered
'animate organism'; for this is equally applicable to Zorba, my old dog,
and Zaner's pooch, Irving. Animals do not possess the lived-body of
ánthrōpos, the physiognomic entity made thematic by the French thinkers
who made the discovery of which I speak. Yet to speak of the body-as-
lived is not to fall into the trap of speaking of the body as '*embodying*

me', as if there is some transcendental or even empirical consciousness inhabiting the physical body, thereby making the physical body a lived-body. This would be to confuse what has to some degree been clarified. Consciousness does not become embodied and (borrowing a Platonic phrase) 'participate' somehow in material Nature. What sense – except a capitulation to the dualism of Cartesians – is there to saying that the body is the besouled or animated embodiment of consciousness? Why speak of incarnate consciousness or consciousness incarnated [*consci-ence-incarnée*]? The metaphysical import of the concept of the body-proper or lived-body is that we can deduce specific categories of corporeity. To be sure, this lived-body is not itself a category; rather, it functions analogously to Being in Aristotle's metaphysics: it is that of which we may articulate categories like (1) uprightness,[22] (2) beholding gaze,[23] and (3) laterality,[24] what I shall take to be the spatial directions indicated when left direction is discriminated from right direction, in opposition.

The lived-body's relation to the world, the all-encompassing other, is not what Husserl meant by *Intentionalität*, but rather indicates a unique relation, *sui generis* and non-causal, a relation to be described not by Brentano's famous 'all consciousness is *consciousness-of*' but by the maxim that 'all human action is a being-able-to'.[25] Merleau-Ponty put it quite succinctly: "La conscience est originairement non pas un 'je pense que' mais un 'je peux'."[26] And this 'I', this 'je', does not have as its referent a transcendental consciousness.

The 'I', as some British philosophers have emphasized, can serve as a grammatical place-holder. Peter Geach astutely notes that 'I' can be used in soliloquy; in this case "there is no question of its referring to anything."[27] The use of 'I' in such contexts is, as Geach says, a "degenerate use", meaning thereby not a term of abuse but a technical term used by mathematicians in describing a certain class of equations. Consider '$0 \cdot x^2 + 3X - 6 = 0$'. When you solve it you ignore the term with 'x^2' in it, no longer using the general rule for solving quadratics, but simply solve '$3X - 6 = 0$'. Merleau-Ponty's phrase, consciousness is an '*I* am able to', makes use of the soliloquistic utterance of 'I' which cancels out much as '$0 \cdot x^2$' does in the equation cited. This is of the utmost importance; the failure to recognize the soliloquistic personal pronoun as merely a grammatical place-holder has led to hopeless searches for the self,[28] for the core of consciousness, and even for a peculiar entity, thus for an only inferred substance or essence. But the '*I* am able to' is only a way of underscoring a remark of Sartre that the body is entirely psychic; that is, the psychic is nothing but a special understanding of the lived-body. Nietzsche put it most succinctly in *Thus Spoke Zarathustra*: "'Body [*Leib*] am I, and soul' – thus speaks the child. And why should one not speak like children? But the awakened and knowing say: body am I entirely, and nothing else; and soul is only a word for something about the body."[29]

III. TOWARD A 'COMPOSITIONAL MATERIALISM'

In "The Perils of Physicalism"[30] Professor Joseph Margolis approaches the doctrine of physicalism – the thesis that a person, with all his psychological attributes, is nothing over and above his body, with all its physical attributes[31] – and suggests that physicalism must confront numerous difficulties. I shall not entertain in any detail the problem of personal identity with which the doctrine of physicalism is usually associated relative to the mind-body identity thesis, but pursue Margolis' suggestion that we may be able to agree, for reasons that are sound, on some form of "coherent materialism that does not entail physicalism or the identity of mental attributes with physical attributes."[32] By the end of his article, Margolis suggests something in the way of a "compositional materialism." This new position in his view does not entail "identity materialism," namely (1) that that of which everything exists is composed of matter and (2) that everything that exists is a physical object,[33] but rather, (1) that although everything is composed of matter (2) not everything that exists is a physical object. This view is exemplified by the thesis, in referring to a work of art, that such a work, a painting, cannot be said to be an abstract entity like a class or universal; similarly, by analogy, a person cannot be said to be a 'soul' or 'psyche', if by those terms we mean an abstract entity, a universal, the product of some sort of transcendental synthesis.

Philosophical phenomenology, founded by Edmund Husserl, has, unlike philosophical anthropology, fundamentally neglected the insights regarding the 'lived-body', phenomenology having had its roots in Kant's transcendental philosophy, having followed quite carefully the main road from Königsberg to Göttingen.[34] In order to appreciate the main line, as it were, of the French antecedents of Margolis' "compositional materialism" (to which I am now somewhat partial), it may not be as inappropriate as it might at first seem, to proceed by responding to the famous exhortation of Otto Leibmann' *"Es muss auf Kant zurückgegangen werden"* [35] – it is necessary to return to Kant!

IV. CATEGORIES OF CORPOREITY

Kant's theory is grounded in the human understanding and thus the objects of experience, including scientific experience, are constituted by the subject in and through synthesizing acts. For him, the transcendental subject constitutes the conditions for the possibility of knowledge. He adds that it had never occurred to Hume that "the understanding might itself, perhaps, through these concepts, be the author of the experience in which its objects are found." [36] Hume, being the kind of empiricist he was, was constrained to derive the categories from experience and its repeated associations.

The philosophical writings of Erwin Straus have seriously challenged this Kantian *Ursprung* or source. In Straus' view, in part due to his earlier medical training and his continued clinical experience, a correctly thought-through philosophical anthropology, having at its foundation a conception of the lived-body, has no need of Kant's epistemological rigour and transcendental machinations. He seeks a new access to and source of the origins of the human world which we can come to know, but he does not appeal to constituting acts of a transcendental subject that is life-less. He begins again, deriving (what I prefer to call) 'categories of corporeity' from lived-experience. Since man, as experiencing being, is a natural creature, Straus and the French phenomenologists have attempted, with success, to reinstate man in nature and once and for all time to "exorcise the persistent Cartesian ghost," as Marjorie Grene has so aptly phrased it. [37]

In *Psychiatry and Philosophy* Straus refers to Kant's first *Critique* and reminds us that Kant "proposes to start from the situation of man. But nowhere," he immediately observes, "does that great work consider us humans as lived-bodies, mobile beings." [38] "Man," Straus continues,

whose mode of knowing is supposedly being investigated, is simply not inserted into earthly space as a live-bodied creature. The sensory sphere, affected by the objects, sits

– so it appears – atop a mobile lived body like a camera mounted on a dolly. A good-natured simpleton, the motor-apparatus carries the burden of reason without realizing – like the legendary Christopher – what it is upholding. It deserves no consideration and receives none.[39]

Thus, challenging Kant's derivation of sensory experience, we may in time reject his transcendental deduction and theory of higher-order synthesis, and flatly complain at the omission of man from his early programme, notwithstanding his remark in the opening passage of the "Transcendental Analytic" in which he begins with a reference to the intuition of objects as given to *man*[40] in a specific fashion. Whereas Kant appeals to the categories of sensibility and the understanding, refusing to appeal to empirical determinants, it is necessary to turn one's attention to the derivation of bodily categories which enable us to constitute the human world in its completeness, a world which is precarious and one which permits infirmity as well as health, death as well as life.

Having attacked Kant's emphasis on the rôle of the categories of the understanding (the rational structure which actively constitutes objects), one should not leave matters at that point. It is important in philosophy, as in medicine, to keep an accurate record and to point out that Kant, though he did not fully develop his own insights, having been swayed by transcendental processes as the source of our knowledge of objects, did in fact mention the mobile being of man and begin to take account of his lived-bodily condition. [I hesitate to speak of man as 'embodied', 'embodied spirit', 'incarnated consciousness' or 'conscious incarnateness' since they all have a Cartesian ring which has, in my view, been adequately refuted and abandoned.]

As early as 1768, some thirteen years prior to the *Kritik der reinen Vernunft* while arguing that absolute space is a "fundamental concept," Kant noted that we can only come to have the concept of 'region' on the basis of the direction or the order of parts as given by one's hands and their relation to the sides of our bodies. He says: "Since the distinct feeling of the right and the left side is of such great necessity to the judgments of the regions, nature has at the same time attached it to the mechanical structure of the human body." He continues, "both sides of the human body, irrespective of their great external similarity, are sufficiently distinguished, by means of clear sensation, leaving aside the differing situation of the internal parts and the perceptible beating of the heart...".[41]

It is of some interest to note that in the original German text Kant always uses *Körper* and never *Leib* in speaking of the human body. He always uses *Körper* for body (12 times); '*menschlichen Körpers*' is employed seven times; bodily space, where movement is crucial, is rendered '*körperlichen Raum*'.[42] And in referring to the heavenly bodies he uses '*Himmelskörper*', clearly accepting the Cartesian *res extensa*, thus conceiving of the human body as analogous to a corpse – just another physical body like a heavenly planet, inert, and moved, if moved at all, from without. All of this serves to support the claim that in Kant's work, it seems, man is not inserted into earthly space as live-bodily creature.

V. LATERALITY

Human ontogenesis, especially in its early years, reveals the fact that differentiation between right and left directions is acquired in stages.[43] If we consider the other directions – up-down, forward-backward (and even inner-outer) – they are constituted by factors other than those necessitating activity on the part of the organism. That is, the factor I shall call '*heftig*', experienced weightedness (accounted for by means of the construct in physics known as gravitational force) is constant, uninterrupted, and already at work prenatally. The priority of down over up is itself constituted not so much by human activity, then, as by the very condition of the physical universe itself experienced as it is for us living on *terra firma*. This is revealed by the universal fact that *ánthrōpos* has to master the upright stance in the earliest months of its life and is even compelled to wage this war against gravity throughout its entire lifetime. The power of the downward toward *terra firma* has existential priority over the upward, vertical direction to the starry skies above which has, perhaps, existential import for the moral law within.[44] And here I leave aside any speculations as to *ánthrōpos* and its existence in non-gravitational fields, a phenomenon no longer only logically possible.

Turning to the forward direction, we see that it takes precedence over the backward, not only because *ánthrōpos* is so constructed as to make walking and running more appropriate in the direction we call 'forward', but also because of the forward directedness of vision, established in and through man's uprightness.

But left and right are originally equal,[45] hence the assumption of the symmetry of our hands and feet and our bodily frame when seen from the dorsal or ventral views. We have to achieve a dominant hand, a dominant eye, a dominant foot and even a dominant hemisphere. It is not always the case, by the way, that a right-handed person is also right-eyed and right-legged. The superiority of the left and left-hand side or, as with the ancient Greeks, the right and right-hand side, may be thought to rest on purely biological or anatomical factors. To be sure, that there is a functional asymmetry of the brain, so-called 'hemispheric dominance', the left cerebral hemisphere frequently being more developed, in some respects, than the right, is a neuroanatomical fact, though whether this is the material cause, or effect, of the superior development of the right hand, is not yet agreed upon by all neuroanatomists and I am unfamiliar with any definitive findings from molecular genetics which argue conclusively for either view. That is, it may well be that the exercise of the hand and fingers leads to the nourishment and subsequent development of that part of the central nervous system which controls these bodily parts. We might eventually have to say that we are left-brained *because* we are right-handed. We too easily have concluded (lacking adequate evidence) that we are right-handed *because* we are left-brained.

It is remarkable that for Aristotle and many after him that *right* was on a par with *above* and *forward*, i.e., there is a lived priority of above to below, forward to backward and *right* to left. However, the right-left discrimination is acquired somewhat later on and, more importantly, in a very different way from the directions above-below and forward-backward. What is important, then, is not whether right-hand side predominates over, is naturally superior to, or is of greater nobility than left (an important thesis for the ancient Greek philosophers),[46] but the fact of the acquired asymmetry in which case one side comes to dominate over or, should I say, operate *against* the other. It is interesting to note that Aristotle held that the *absence* of any marked distinction between right and left in some species is due to their imperfect and even 'deformed' status.[47] And Kant understood the *feeling* between the left and right to be pregiven whereas in fact it is necessarily acquired.

The left-right distinction is at first an opposing, a pointing toward the left, a pointing toward the right. So it is not correctly described as left *and* right, but rather left *versus* right. The pointing is one of opposite directions; eventually one leads and the other is made subordinate.

Neurologically, of course, we can speak of cerebral 'dominance' and 'non-dominance' wherein we mean by non-dominance something analogous to 'recessive' in the taxonomy of genetic factors. Perhaps 'ordinate' and 'subordinate' are more appropriate terms to describe the function of the cerebral hemispheres and sidedness. Just as the lateral functions of our arms, hands, and fingers is constituted by activity (having already been set free by the upright stance), so too the cerebrum and its outer bark, the cortex, is differentiated; this is well known from the plethora of neurological findings determined upon post mortem examination. So whereas right *versus* left divides the human frame symmetrically, given the normal conditions of development of *ánthrōpos*, so too is the asymmetry of left *versus* right acquired through function and activity.

The notion of hemispheric dominance is of little use in and of itself in taking us beyond the necessary condition of human activity, whether involving the motility of our bodily parts or even our ability to generate language. As a matter of fact, it may well be more correct to say that hemispheric dominance is brought about through favoring right or left rather than presume that a prior dominance of one or the other hemisphere is the *cause* of lateral dominance or sidedness. I have always thought that Aristotle was in error in criticising Anaxagoras when in *Parts of Animals* he says: "Now it is the opinion of Anaxagoras that the possession of these hands is the cause of man being of all animals the most intelligent. But it is more rational to suppose that his endowment with hands is the consequence rather than the cause of his superior intelligence."[48] Anaxagoras, who Aristotle had said was "a sane man among babblers,"[49] had already seen that the actual use of the hands is itself the *material* cause of man's intelligence. The preferred position in our time is, of course, to argue that the brain of man has been differentiated through an evolving process, such that a greater portion of its composition is the necessary condition for the continuous functioning of the human hands and fingers.

VI. NEUROPATHOLOGY: THE GERSTMANN SYNDROME

In order to motivate philosophical reflection and to indicate its appropriate course, the domain of pathology as viewed by the medical subspecialties of neurology and psychiatry has proved itself invaluable to some philosophers. Kurt Goldstein, whose reputation in the domain of the aphasic disorders is renowned,[50] once remarked in his autobiographical essay that he viewed himself as devoting his efforts "to solving the most complicated and essential epistemological problem of biology and medicine": namely, how an apparently inconspicuous symptom, such as loss of ability to find a word under special conditions, (e.g.) brain damage, could be masked by a patient.[51]

It is now a sound methodological principle to attend to disabilities (or failures) in order to shed light on normal performances which we always take for granted. In everyday life we take it for granted that a normal person will be able to add and subtract the simplest sums, write correctly the simplest words, remain oriented as he moves around his house or neighborhood after years of residence in these places, and surely tell on which finger of either hand he has been touched. It is well known that, subsequent to certain pathological manifestations (such as a specific insult or organic lesion, and especially those well-defined ones established by the neurologists through correlation methods subsequent to autopsy), that certain patients experience various types of spatial disorientation; certain disturbances of spatial orientation themselves count among neurological symptoms.

In 1924 a Viennese neurologist, Josef Gerstmann, first described a symptom complex which appeared not infrequently in patients suffering from a cerebral lesion located in the transitional area of the lower parietal and the middle occipital convolution, namely, in that part which corresponds to the angular gyrus in its transition to the second occipital convolution of the *dominant* hemisphere. This symptom complex, according to Gerstmann, consisted in a tetrad: (1) finger agnosia, (2) disorientation of right and left sides, (3) agraphia and (4) acalculia.[52] Gerstmann maintained (even in a posthumously published rejoinder to Critchley) that he had established the anatomic correlate of the symptom complex, which was definable as a circumscribed cerebral syndrome. By 1940 Gerstmann considered the localizing value of the syndrome as "proved."[53]

As a neurologist, Gerstmann was most concerned with firmly establishing the relation between the symptom complex and the specificity of the damage in the designated area of the parietal lobe of the dominant hemisphere. The benefit of such knowledge is surely invaluable to the physician in all diagnoses in which he hopes to infer from specific symptoms revealed by the patient the organic damage or impairment in the brain and/or central nervous system. Yet in spite of the establishment of the uniform localization of a focal lesion Gerstmann was not satisfied that he had fully understood precisely what "consolidates" the individual (but not isolated) elements of the syndrome as a basic disturbance, as a fundamental unit. He hoped to discover why it is that the tetrad of symptoms appears as a cluster, why the four behavioral deficits tend to associate as they in fact seemed to do.[54]

The philosopher who attends to the phenomena of pathology and, say, the Gerstmann symptom complex in particular, is not especially interested in the neurological findings, i.e., the exact location of the lesion in the brain, for it tends to be philosophically irrelevant whether a specific locale is highly correlated with the behavioral deficits or whether, as some may hold, the entire brain is involved in the adequate account of all deficits in performance. Moreover, it has not been until quite recently that philosophers have found pathological findings philosophically relevant at all. But philosophers who have a more concrete and empirical bent and who distrust transcendental inquiries have turned to pathological phenomena for clues to the norm of the life of *ánthrōpos*.

It is of interest to note that classical neurology always had a chapter dedicated to spatial disorientation, and Gerstmann's syndrome was one of those designated by different authors as fundamentally an impairment of the patient's concept of space, of his or her capacity to apprehend figure-ground relationships, or of the patient's sense of direction in space. But such conclusions are in fact too general. For the tetrad of symptoms constituting Gerstmann's syndrome can all be understood as a disturbance of laterality – the directions left and right – a breakdown of the acquired opposition of right *versus* left. It seems to be peculiar to *ánthrōpos* that only the human species constitutes the asymmetry required before he can (1) respond with accuracy to requests to locate fingers touched,[55] (2) write the simplest words as he did before the cerebral accident or (3) add and subtract arithmetic sums on paper as was easily accomplished prior to the insult in the form of a specified parietal injury.[56]

The acquired distinction between left-side and right-side is not explicable as a mere linguistic habit learned by repeated associations, but is a consequence of human action enabling the most complex achievements even beyond adding and subtracting, multiplying and dividing sums. The acquired opposition of the lateral directions (I shall maintain) each pointing in opposite directions is the condition of the possibility of spatial orientation and, *mutatis mutandis*, perhaps all so-called 'cognitive' processes, produced by the dominance of one lateral direction over and against the other.

VII. ACALCULIA AND POSITION IN LATERAL SPACE

One of the components of the tetrad noted by Gerstmann is, as I have already mentioned, acalculia or discalculia. First coined by Henschen in 1919,[57] acalculia is the name he gave to a disturbance of calculating, produced by a focal lesion of the brain. It is important to distinguish (1) acalculia in the broad sense in which the patient has difficulty writing numerals on verbal command, or copying the numerals from a written illustration, or even reading aloud the word 'three'[58] when shown '3'[59] on a card, from (2) acalculia in the narrow yet more instructive sense in which the patient fails to 'reckon', i.e., add and subtract, multiply and divide. Acalculia in this narrow sense has remained of interest to some, thanks to Gerstmann's writings and Macdonald Critchley's *Parietal Lobes*.[60] This impairment is quite frequently associated with finger agnosia, associated, that is, with the parietal lobe lesion described by Gerstmann as early as 1924. I shall not concern myself with the broad sense of acalculia then, but with the failure to reckon, on the part of those patients, having already been quite competent at reckoning prior to the brain lesion due, usually, to a cerebral accident of some sort.

In 1644 J. Bulwer wrote the following:

To begin with the first finger of the left hand, and to tell on to the last finger of the right, is the natural and simple way of *numbring* and *computation*: for, all men use to count forwards till they come to that number of their *Fingers*.... Hence some have called man a naturall Arithmetician, and the only creature that could reckon and understand the mistique laws of numbers, because he alone hath reason, which is the spring of arithmeticall account: nay that divine Philosopher doth draw the line of men's understanding from this computing faculty of his soule, affirming that therefore he excells all creatures in wisdome, because he can account....[61]

Rather than accept blindly the claim that the faculty of the human soul is the condition which enables man to reckon and calculate, it is perhaps more rewarding to study the processes involved in doing simple arithmetic and to do so before offering an account of acalculia other than determining the locus of the lesion in the brain, which has no particular philosophical significance. What *is* of great significance are the features of the Hindu system for the notation of numbers, which proceeds by ones, tens, tens of tens (hundreds), etc. Not only does this system (1) limit the number of different symbols to ten, however large the numbers referred to in the problem and (2) use zero to indicate the absence of a power, but (3) the value of a digit is determined according to its *position* within a number-notation. The plural cipher notation, therefore, proceeds from right to left and the value of a digit within a number notation is counted from the right, and its place corresponds to the power of the base (commonly the base 10). The '0' is like the 'I' in 'I am my body' or the 'Je' in 'Je peux' – a place-filler. The '0' also enables us to add units to units, tens to tens, hundreds to hundreds, etc. Thus we use columns for convenience, but this necessitates being able to keep columns distinct from one another while at the same time working our way from right to left, having written the ciphers or notations left to right!

In adding and subtracting we have to 'carry over' and 'borrow from', respectively, every ten units of one denomination. We 'carry to' *and* 'borrow from' the left; we are forced to regroup, a process requiring clarity as to the lateral direction – left side *versus* right side. So *in the base* 10, $11 + 2$ is $10 + 3$ and not $7 + 6$. Subjects who fail to carry out such procedures may, however, still reckon well 'mentally'. But ask yourself whether you reckon 'in your head' as you do on paper. I think not. That is, many, though not all persons, if asked to add 76 and 27, find that they do something like the following: "Seventy-six and twenty is ninety-six and seven more has to be added on; four more gets me to one-hundred and three to a hundred and three." During this procedure (which Piaget has articulated in great detail with astounding results),[62] one does not have to utilize lateral directionality, right-to-left. Rather, one may deal with chunks, wholes – even ones, to be sure; and then with the remainder under ten, one can consider these digits (from which the word 'finger' is derived) very rapidly in a counting-off fashion, picking up the four to get to the hundred; then, saying "a-hundred-tack-on-the-remaining-three," I cluster the new whole. I submit this as partial evidence (which may generate empirically testable hypotheses) to justify the claim that acalculia in the narrow sense is primarily a disorder of *lateral* orientation which may still permit some subjects to calculate or reckon in their heads with some precision. For calculating and reckoning on paper require a stable orientation in the lateral direction, established in us by the action of arms, hands, and fingers; this sidedness orientation is, as I see it, the missing ground of the logical category of *position* in Aristotle and *reciprocity* in Kant.

A survey of Aristotle's ten categories or predicables[63] reveals the often neglected "position" [*situs*] of which even he says very little in Chapter Nine of *Categoriae*, except that he already spoke of this category when he dealt with the category of relation.[64] Thus, 'to the left of' and 'to the right of' are relations or predicates of position. In Aristotle's view, the distinction between right and left in man is what enables him to realize a change of place[65] (place itself being another category [*ubi*]); but in *De Incessu Animalium* he remarks that "the right is that from which change of position naturally begins, the opposite which naturally depends upon this is the left."[66]

Kant, modifying Aristotle's categories for his own purposes, speaks of 'reciprocity' as the category which enables the constitution of space. And in the Kantian system this category is the condition of the possibility of laterality. Kant remarks in this context, as is well-known, that he follows Aristotle's logic, list of categories, and table of judgments, since "our primary purpose is the same as his although widely diverging from it in manner of execution."[67] Yet Kant failed to see the full implication of the fact that such a category is not a product of transcendental consciousness, but rather the lived-body of *ánthrōpos*; laterality is quite appropriately viewed as a category of corporeity, one among others to be sure, and requiring the same effort at discovery as did Aristotle's logical categories.

The study of acalculia (I submit), first noted by physicians who were astute observers, has important philosophical, i.e., metaphysical, implications. The lived-body and not the physical body is at the boundary or interface of philosophy and medicine; this is the crucial boundary condition of these two disciplines. We can move quite readily, then, from the pathological evidence of a parietal injury to neuro-psychology and neurology, then on to the pathological phenomena of acalculia and end with philosophical relevance and the articulation of a categorial scheme having its source in the lived- rather than physical-body, a distinction which is itself a product of philosophical reflection.

VIII. "ES MUSS AUF KANT ZURÜCKGEGANGEN WERDEN!"

It is somewhat paradoxical that the most eminent of philosophers, having based his system on transcendental subjectivity, has in fact quite obliquely suggested a category of corporeity, when his *magnum opus* had as a central thesis the establishment of categories of sensibility and categories of the understanding. After all, the very *Lehrstück* of his philosophy is that the supreme principle of the constitution of all knowledge or cognition is the activity of the experiencing subject itself. The categories denoted functions that pertain to the experient subject and derive from what a subject knows of itself. For Kant, we may recall, a 'category' means "the function of synthesizing into a definite unity the manifold of items of a given intuition."[68] But Kant's categories, as we have seen, synthesize from a *lifeless* perspective; it is the motility of our newly-freed arms and hands as well as the action of our corporeity that, in living, constitutes the unity of our experience and makes many, if not all, so-called higher-order "cognitive" achievements possible.

Yet, once again – following sound medical practice – we must keep accurate records and not distort the data. Another look at the Kantian *corpus* reveals that Kant had a second chance to redeem himself for his original sin: In 1786, some five years after the appearance of the first edition of the first *Kritik*, Kant published *Was heisst: Sich im Denken orientieren?*[69] While considering geographical orientation, Kant refers to the fact that in having a right and a left hand we can orient ourselves geographically. Kant's words are truly revolutionary:

To orient one's self in the strict sense of the word means to find from one given direction in the world (one of the four into which we divide the horizon), the others, especially the east. If I see the sun in the sky and know that it is now noon, I know how to find south, west, north, and east. But for this I certainly need the feeling of a distinction in my own person, that between my right and left hand. I call it a feeling because the two sides in intuition show no externally noticeable difference. Without the capacity to distinguish between motion from left to right and that in the opposite direction in describing a circle, in spite of the absence of any difference in the objects, I would be unable to determine a priori any difference in the position of objects; I would not know whether to put west to the left or right of the south point of the horizon so as to know how to complete the circle from north through east to the south. Thus I orient myself geographically by all the ob-

jective data for the sky only by virtue of a subjective ground of distinction, and if someday a miracle occurred whereby the direction of all the stars changed from east to west
but preserved the same patterns and position on the next starry night no human eye would
notice the slightest change, and even the astronomer, if he attended to what he merely
saw and not what he at the same time felt, would be inevitably disoriented. But the ability
to distinguish by feeling between the right and the left hand, which is implanted by nature
and made familiar by frequent use, would come naturally to his help, and if he once viewed
the pole star, he not only would notice the change but would orient himself regardless
of it....[70]

Kant, then, can in fact be credited with acknowledging man's lived, motile
body, even though he employs this notion to sustain the thesis that we
intuit and construct space through "a subjective ground of distinction."
Kant roots our construction of space in the fact of our having a "feeling"
of difference between our right and our left hand. By 1786 Kant abandoned the misleading model of man as a camera mounted on an unfeeling
dolly. For the dolly now experiences its left as distinct from its right hand
by means of a subjective ground or feeling of itself as motile and alive.

IX. THE LIVED-BODY AS CATALYTIC AGENT

I have referred to the lived-body as a catalytic agent for four reasons:
First, the lived-body (and for that matter the lived-body as a mere physical body) does not alter in any way by introducing a new categorial
metaphysic (like the one which I have suggested) for its comprehension
and for the comprehension of certain kinds of infirmities; only our attitude toward it is modified. Second, the lived-body is itself a *tertium
quid* and may be the subject matter of two disciplines, but distinct from
both. Third, if physicians and philosophers can come to agree that the
lived-body is the meeting point of their respective disciplines, then man's
corporeity, as understood through a categorial metaphysic, will serve to
accelerate the reaction at the interface of both Medicine and Philosophy;
this will have a cultural impact of some import in addition to the con-
nexus made by medical ethics and the philosophy of science in transition
to a philosophy of medicine. Finally, the totally conceptualized lived-
body is not itself a sufficient account of *ánthrōpos*, since 'personhood' too
must be accounted for and, at the very least, a clarification of that concept
requires an explication of human agency, action, conduct, and even the
bodily conditions of the possibility of transcendence.

I have not discussed the self, personhood, or human agency; that will require another time and additional space. My title, in which I refer to a catalytic agent, can only serve to titillate, but I would intend, if time were not so limited, to explore the notions of self, personhood, and human agency, since those features too are, I believe, conditioned by structures of our bodily physiognomy. But that would be the last chapter of my thus far unwritten treatise, and for its completion I shall have to plead for a few more years.

X. EPILOGUE

In my introductory remarks I included reference to Plotinus' life regarding which we have, still preserved, Porphyry's account. He tells us that

Plotinus was often distressed by an intestinal complaint, but declined clysters,[71] pronouncing the use of such remedies unbecoming in an elderly man: in the same way he refused such medicaments as contain any substance taken from wild beasts or reptiles.... When the terrible epidemic carried off his masseurs he renounced all such treatment: in a short while he contracted malign diphtheria. During the time I was about him there was no sign of any such malady, but after I sailed for Sicily the condition grew acute: his intimate, Eustochius, who was with him till his death, told me, on my return to Rome, that he became hoarse, so that his voice quite lost its clear sonorous note, his sight grew dim and ulcers formed on his hands and feet.[72]

As we read on we discover that the intimate companion of Plotinus, Eustochius, who was with him until his death, eventually "consecrated himself exclusively to Plotinus' system and became a veritable philosopher."[73] Such discipleship served in some small measure, in addition to Platonism, Neoplatonism, and their periodic revivals, to sustain the banner of spiritualistic metaphysics, the abandonment of which I have to some degree tried to argue for in the preceding pages.

Today the impetus of the spiritualistic impulse still poses difficulties for the contemporary physician, who, if it has not yet been made perfectly clear, is joined in arms by many contemporary philosophers, especially of the Continental persuasion, in rejecting spiritualistic metaphysics and speculative excesses. Finally, for the purpose of maintaining historical accuracy, we ought not to forget that Plotinus' metaphysics was shared by others during his own time, even "veritable philosophers," like Eustochius of Alexandria, who, I almost failed to mention, was personal physician to Plotinus.

University of Connecticut Health Center,
Farmington, Connecticut

NOTES

* The initial research and preparation of this paper was accomplished while a Fellow of the Institute on Human Values in Medicine under grant number EH-7278-73-104 of the National Endowment for the Humanities to the Society for Health and Human Values. I extend my gratitude to the Institute, Society and Endowment for their support and sponsorship. I also wish to thank the Department of Medicine, especially Dr. George Wolf, of the University of Vermont College of Medicine for hosting me while in residence throughout the summer, 1973.
[1] *Cartesian Meditations: An Introduction to Phenomenology*, trans. by Dorion Cairns (Hague: Martinus Nijhoff, 1960), p. 139. In his analysis of the constituting of intersubjectivity in the "Fifth Meditation," Husserl remarks that what he means by metaphysics is "anything but metaphysics in the customary sense" [Aber nichts weniger als Metaphysik im gewohnten Sinne ist hier in Frage]; for the sense of metaphysics has become "historisch entartete." See the original text, *Cartesianische Meditationen und Pariser Vorträge*, 1929, ed. by S. Strasser, *Husserliana*, Band I (Hague: Martinus Nijhoff, 1950), p. 166.
[2] "Euthanasia," Editorial, *Journal of the American Medical Association* **218** (Oct. 11, 1971), 249. Negative euthanasia is currently defined as the "planned omission of therapies that probably would prolong life."
[3] I shall follow the current convention and use upper case to indicate that 'Medicine' includes all the health science disciplines and is not restricted in its reference to the specific province of physicians. For an excellent presentation of non-degenerate metaphysics see D. F. Pears (ed.), *The Nature of Metaphysics* (London: Macmillan, 1957).
[4] Husserl, *Cartesian Meditations*. Husserl maintains that his "monodological results are metaphysical" and he hoped to return to the sense "with which metaphysics, as 'First Philosophy', was instituted originally." What is perverse and to be rejected is "*alle spekulativen Überschwänglichkeiten.*"
[5] Plotinus, *The Enneads*, trans. by Stephen MacKenna (London: Farber and Farber, 1956), p. 4. MacKenna translates Porphyry's reference to Tractate Eight of the Fourth Ennead as "On the Descent of the Soul into Bodies"; this is later translated as "The Soul's Descent Into Body" when used as the rubric for IV, 8 on page 357. Plotinus' commentator, Porphyry, may have indicated that Plotinus ascribed to a doctrine of transmigration of Soul or even a doctrine of metempsychosis or metamorphosis. That is, the Soul may, throughout time, become incarnated in *various* bodily forms, human as well as nonhuman. Whether or not this was Plotinus' view we need not consider here. Suffice to say that for Plotinus the Soul can disengage from its bodily matrix (though not in any spatial sense as is evidenced by V, 1, 10) and re-engage that particular body when it 'descends' from some supercelestial topos.
[6] *Ibid.*, IV, 8, 1.
[7] *Ibid.*, "The Immortality of the Soul," IV, 7, 1. It is of interest to note that in the "Daily Prayer of a Physician," attributed to Moses Maimonides, the human body is referred to as "envelope of the immortal soul." See Fred Rosner, "Physician's Prayer Attributed to Moses Maimonides," *Bulletin of the History of Medicine* **45** (Sept.-Oct., 1967), 451.
[8] *Ibid.*, 'On the Life of Plotinus and the Arrangement of his Work," p. 1.
[9] See H. Tristram Engelhardt, Jr., "The Concepts of Health and Disease," in this volume, p. 125. It is necessary to distinguish disease nosologies from disease processes; more importantly, though strange to the ears perhaps, there is the view that there are no diseases in Nature, only sick people.
[10] I use the term 'the body' to signify the human, living body of a person. It will not be

used to refer to the human corpse or cadaver. See Stephen Toulmin, "Concepts of Function and Mechanism in Medicine and Medical Science: Hommage à Claude Bernard," in this volume, p. 51. Also see S. Toulmin and J. Goodfield, "Claude Bernard's New Method," in *The Architecture of Matter* (New York: Harper and Row, 1962), pp. 331–336.

[11] The term 'body image' is used in a wide variety of ways in philosophical and psychological literature; it often serves to conceal a theoretical position left unexplicated. See Paul Schilder, *Image and Appearance of the Human Body* (New York: International Universities Press, 1950); Seymour Fisher and Sidney E. Cleveland, *Body Image and Personality* (New York: Dover Publications, 1958); Warren Gorman, *Body Image and the Image of the Brain* (St. Louis, Missouri: Warren H. Green, 1969).

[12] *Man a Machine*, French-English edition, with philosophical and historical notes by G. C. Bussey (La Salle, Illinois: Open Court Publishing Co., 1961). The original text was first published in 1748. I am personally indebted to Professor Stephen Toulmin for introducing me to the work of la Mettrie. In 1966, at the University of Colorado, Professor Toulmin presented a paper in which he explicated the confusion among philosophers as to the significance of la Mettrie's *L'Homme Machine*. The paper, "Neuroscience and Human Understanding," was subsequently published in *The Neurosciences: A Study Program Planned and Edited by Gardner C. Quarton, T. Melnechuk and F. O. Schmitt* (New York: Rockefeller University Press, 1967), pp. 822–832. Also see *The Architecture of Matter*, "The Animal Machine," pp. 307–337. This chapter is a splendid analysis of la Mettrie's contribution to medicine and philosophy as well as an explication of various theories of vitality.

[13] "Neuroscience and Human Understanding," p. 823.

[14] *Ibid.*, p. 824 *et passim*.

[15] *The Architecture of Matter*, p. 318.

[16] See Rudolf Bultmann, *Theology of the New Testament: I*, trans. by Kendrich Grobel, pp. 192–203, "Soma" in Paul's Theology. Also see N. Schneemann, "Leib, Existenz, Theologie," *Confinia Psychiatrica* 15 (1972), 125–148.

[17] We do not intend to suggest that the anatomist and physiologist do not carry out their studies with respect to the general concern of the processes of living beings. But the biomedical scientist does not, methodologically, always have the opportunity to study structures and functions except by way of the cadaver, the stained slide, the laboratory experiment. Their principles are biological and not perhaps applicable to other levels of organization of matter. See Michael Polanyi, "Life Transcending Physics and Chemistry," *Chemical and Engineering News* 45 (August 21, 1967), 55–66; *The Tacit Dimension* (New York: Doubleday, 1966), pp. 87–91; see Toulmin, *Architecture of Matter*, pp. 318ff. Here la Mettrie's notion of 'organization' is presented as an earlier and sketchy picture; yet it led to the claim that the science of organized bodies should be treated in a manner quite distinct from the sciences whose subject matter is unorganized bodies.

[18] (New York: Dell Publishing Co., Delta Book, 1964), p. 92.

[19] J. H. Van Den Berg, "The Human Body and the Significance of Human Movement: A Phenomenological Study," *Philosophy and Phenomenological Research* 13 (December, 1952), 169–170.

[20] *The Problem of Embodiment* (Hague: Martinus Nijhoff, 1964), p. 23. See Dorion Cairns, *Guide for Translating Husserl* (Hague: Martinus Nijhoff, 1973), p. 79.

[21] Zaner, *The Problem of Embodiment*, p. 23. Zaner cites Edmund Husserl's *Cartesian Meditations*, trans. by D. Cairns (Hague: Martinus Nijhoff, 1960), p. 97.

[22] See Erwin Straus, "Die aufrechte Haltung: eine anthropologische Studie," *Monatsschrift für Psychiatrie und Neurologie* (Basel/New York: S. Karger, 1949), Band 117,

Heft 4/5/6. The English translation "The Upright Posture," appears in Straus, *Phenomenological Psychology* (New York: Basic Books, 1966), pp. 137–165.

[23] See E. Straus, "Zum Sehen Geboren, Zum Schauen Bestellt: Betrachtungen zur 'Aufrechten Haltung'," *Werden und Handeln*, herausgegebene von Eckart Wiesenhutter (Stuttgart: Hippokrates-Verlag, 1963), pp. 44–73. The English translation, "Born to See, Bound to Behold: Reflections on the Function of Upright Posture in the Esthetic Attitude," appears in *The Philosophy of the Body*, ed. by S. F. Spicker (New York: Quadrangle Books, 1973), pp. 334–361.

[24] See E. Straus, "Über Anosognosie," *Jahrbuch für Psychologie und Psychotherapie*, 11 Jahrgang, Heft 1 (Freiburg/München: Karl Alber Verlag, 1964). The English translation "Anosognosia," appears in *Phenomenology of Will and Action* (Pittsburgh, Pa.: Duquesne University Press, 1967), especially pp. 116–126. I am deeply indebted to Dr. Straus for his insights on the topic of laterality and left-right disturbances due to insult and organic injury. Our frequent conversations on this topic enabled me to appreciate the vast domain of work to be done in the area of a philosophy of corporeity.

[25] See Maurice Merleau-Ponty, *Phénoménologie de la perception* (Paris: Librairie Gallimard, 1945), p. 160.

[26] I render this, "Consciousness is, fundamentally, not an 'I think that' but an 'I am able to'."

[27] "Reincarnation," in *God and the Soul* (London: Routledge and Kegan Paul, 1969), p. 8.

[28] I say 'hopeless searches' not in the sense that the concept of self has no significance or referent, but in the sense that, as Zaner has argued, the search was generally misguided in seeking either a simple *ens* or an abstract entity having a peculiar ontological status.

[29] See *The Portable Nietzsche*, trans. by Walter Kaufmann (New York: Viking Press, 1967), the English edition, First Part, "On the Despisers of the Body [des Leibes]," p. 146. The original text reads: "'Leib bin ich und Seele' – so redet das Kind. Und warum sollte man nicht wie die Kinder reden? Aber der Erwachte, der Wissende sagt: Leib bin ich ganz und gar, und nichts ausserdem; und Seele ist nur ein Wort für ein Etwas am Leibe." See my "Introduction," *The Philosophy of the Body*, pp. 3–23.

[30] *Mind* 32 (October, 1973), 566–578.

[31] *Ibid.*, p. 566. Also see Thomas Nagel, "Physicalism," *The Philosophical Review* 74 (1965), 339–356.

[32] *Ibid.*, p. 575.

[33] *Ibid.* I am intentionally, yet only partially, misconstruing Margolis' intention.

[34] Husserl's career began as *Privatdozent* at the University of Halle (1887–1901), continued at Göttingen (1901–1916) and ended in Freiburg im Breisgau, where he died in 1938. Herbert Spiegelberg remarks that it was while at Göttingen that Husserl developed his "pure phenomenology as the universal foundation of philosophy and science, which takes shape around 1906 and soon leads not only to the formulation of a new transcendentalism but of a characteristic phenomenological idealism, whose increasing radicalization is the main theme of Husserl's period in Freiburg." See *The Phenomenological Movement: A Historical Introduction*, I. (Hague: Martinus Nijhoff, 1960), p. 74.

[35] See *Kant und die Epigonen* (Cannstatt: Verlag von Emil Geiger, 1865), p. 215.

[36] *Ibid.*, Immanuel Kant, *Critique of Pure Reason*, trans. by Norman K. Smith (London: Macmillan, 1958), p. 127 (A-94, B-127). *Vide* A-79–94; B-106–127.

[37] *Approaches to a Philosophical Biology* (New York/London: Basic Books, 1965), p. 194.

[38] (Berlin, Heidelberg, New York: Springer Verlag, 1969), p. 41.

[39] *Ibid.*

[40] Kant, *Critique of Pure Reason*, p. 65 (B-34).

[41] "Concerning the Ultimate Foundation of the Differentiation of Regions in Space," in *Kant: Selected Pre-Critical Writings and Correspondence with Beck*, trans. by G. B. Kerferd and D. E. Walford (England: Manchester University Press, 1968), p. 40. In this work Kant distinguishes the concept of 'position' of objects in space from that of 'region' (the condition for a system of positions, what he calls 'absolute' or 'original' space).

[42] *Ibid.*, "Von dem ersten Grunde des Unterschiedes der Gegenden im Raume," *Sämmtliche Werke*, II, pp. 385–391, *et passim*.

[43] Discussion with Erwin Straus, April 12, 1974, Tampa, Florida.

[44] On the relation that obtains between the physiognomy of man's upright posture and his ability to constitute a moral life, see E. Straus, "The Upright Posture." Also see N. Schneeman, "Leib, Existenz, Theologie."

[45] For a discussion of developmental dimensions of sidedness see Henry Hecaen and J. de Ajuriaguerra, *Left-Handedness*, trans. by Eric Ponder (New York/London: Grune & Stratton, 1964), pp. 1–4.

[46] See G. E. R. Lloyd, "Right and Left in Greek Philosophy," *Journal of Hellenic Studies* **82** (1962), 65–66. See *De Partibus Animalium*, 3. 3. 665a 22–25; 9. 671b 29–35; 4. 666b 3–10.

[47] *Ibid.*, p. 65. See *De Incessu Animalium* 706a 21ff.

[48] Aristotle, *De Partibus Animalium: Book I*, trans. by William Ogle (Chicago: University of Chicago Press, 1962), 4. 10. 687a 7–23. Hereafter all references are to *Works of Aristotle*, trans. and ed. by W. D. Ross, 12 vols. (Oxford: Clarendon Press, 1908–52).

[49] *Metaphysica*, Book A, Ch. 3, 984b 18.

[50] See "Bibliography of the Published Writings of Kurt Goldstein," prepared by Dr. Joseph Meiers, printed in *The Reach of Mind: Essays in Memory of Kurt Goldstein*, ed. by Marianne Simmel (New York: Springer Publishing Co., 1968), pp. 271–295.

[51] "Kurt Goldstein," in *A History of Psychology in Autobiography*, ed. by E. G. Boring and G. Lindzey, Vol. 5 (New York: Appleton-Century-Crofts, 1967), p. 153.

[52] "Fingeragnosie: Eine umschriebene Störung der Orientierung am eigenen Körper," *Wiener klinische Wochenschrift* **37** (1924), 1010–1012.

[53] Josef Gerstmann, "Syndrome of Finger Agnosia, Disorientation for Right and Left, Agraphia and Acalculia," *Archives of Neurology and Psychiatry* **44** (1940), 398–408.

[54] *Ibid.*, p. 404.

[55] *Ibid.*, p. 399 *et passim*. Also see Macdonald Critchley, *The Parietal Lobes* (New York: Hafner Publishing Co., 1953), pp. 206–207.

[56] Critchley, *The Parietal Lobes*, p. 208. Also see Erich Guttmann, "Congenital Arithmetic Disability and Acalculia (Henschen)," *British Journal of Medical Psychology* **15-16** (1935–37), 16–35.

[57] F. Grewel, "Acalculia," *Brain* **75** (1952), 397. The reference is to S. E. Henschen, "Ueber Sprach-, Musik- und Rechenmechanismen und ihre Lokalisation im Gehirn," *Zeitschrift ges. Neurologie und Psychiatrie* **52** (1919), 273.

[58] 'Numeration' is often used to refer to the representation of numeral quantities by word, e.g., 'three'.

[59] 'Notation' is often used to refer to the representation of numeral quantities by signs, e.g., '3'.

[60] Critchley, *The Parietal Lobes*, pp. 203–224. Critchley modified his view of Gerstmann's claim to the discovery of the syndrome bearing his name in "The Enigma of Gerstmann's Syndrome," *Brain* **89** (June 1966), 183–198. In this article Critchley maintains that the tetrad of disabilities are in fact "a most heterogeneous collection of disabilities" (p. 186) and concludes that he is "unconvinced as to the *bona fides* of Gerstmann's syndrome" (p. 196). In a posthumous publication Josef Gerstmann (who died on March 23, 1969)

presented a rejoinder to Critchley's criticism of his earlier discovery. The arguments and counterarguments are of interest in their own right and even serve to suggest that the concept of 'syndrome' in medicine requires some analysis. See "Some Posthumous Notes on the Gerstmann Syndrome," in *Wiener Zeitschrift für Nervenheilkunde und deren Grenzgebiete* (Wien/New York: Springer Verlag, 1970), Bd. XXVIII, Heft 1. I have tried to address my remarks to Critchley's question when he says, "But to align the Gerstmann syndrome with a basic defect in spatial conception raises difficult questions such as why only lateral dimensions are confused. Scarcely," he adds, "if ever does the patient show uncertainty in vertical orientation, or in such concepts as *in* and *out*; *far* and *near*; *above* and *below*." See *Brain* **89** (June 1966), 191. In my view, the answer to this question lies in the phenomenon of right *versus* left – not right *and* left – and the acquiring of asymmetry in the lateral directions which does not take place in other bodily-spatial directions or dimensions. *Vide* Henry Hecaen and J. de Ajuriaguerra, *Left-Handedness*, pp. 118–121.

61 Critchley, *Parietal Lobes*, p. 203. This is taken from J. Bulwer, *Chirologia* (London, 1644), and cited by Critchley.

62 See Jean Piaget, *The Child's Conception of Number* (New York: Humanities Press, 1952); also "La gènese du nombre chez l'enfant" in J. Piaget, B. Boscher and A. Chatelet, *Initiation au calcul* (Paris: Bourrelier, 1956), pp. 5–28.

63 *Categoriae*, 4. 1b 25–30; also see Kant, *Prolegomena*, Section 39.

64 *Ibid.*, 9. 11b 7–10.

65 *De Incessu Animalium*, 4. 705b 14–15.

66 *Ibid.*, 705b 17–21.

67 Kant, *Critique of Pure Reason*, A-80, B-106.

68 See James Ward, "Immanuel Kant," in *Essays in Philosophy* (Cambridge: Cambridge University Press, 1927), p. 331.

69 *Kants Werke* (Berlin: Walter de Gruyter, 1968), Bd. VIII, pp. 134–135.

70 "What Does it Mean to Orient Oneself in Thinking?" *The Philosophy of the Body*, ed. by S. F. Spicker, pp. 96–97.

71 From the Greek 'klystēr' (to wash out): enema; a rectal injection of water, gas or other fluid.

72 Plotinus, *The Enneads*, "Porphyry's Life," p. 1, Section 2.

73 *Ibid.*, p. 6, Section 7.

ANDRÉ SCHUWER

COMMENTS ON "THE LIVED-BODY AS CATALYTIC AGENT"

Professor Spicker's "The Lived-Body As Catalytic Agent" is a good example of a reaction at the interface of medicine and philosophy. Spicker
made the lived-body [*le corps propre*] thematic as the meeting point of
the disciplines of philosophers and physicians. The category of laterality
is suggested as a means to comprehend the lived-body and to comprehend
certain infirmities. We will articulate our discussion in four points.

(1) Spicker reacts at the very beginning of his paper against the "spiritualist extravagance of the Occidental tradition" (p. 181). It is worth noting that his illustrative traditional metaphysician is not Descartes but
Plotinus – so that the mind-body dualism is located over fourteen centuries
earlier in the history of philosophy. But we have first to ask here, what
does Spicker understand by mind-body dualism? Spicker, it seems to us,
is construing something rather more extended than what is normally
laid to the door of Descartes. For he does not simply reject the Cartesian
notion that only the *res cogitans* is in principle indubitable (and its accompanying claim that only knowledge acquired through it, and not
knowledge acquired through the *res extensa* – including not simply sense-
organs but the imagination also – is indubitable); but beyond this,
Spicker is questioning any application, for example, of the expressions
"embodiment of consciousness," or "incarnate consciousness" or, again,
"animated organism." When Spicker writes: "What sense – except a capitulation to the dualism of Cartesians – is there to saying that the body
is the besouled or animate embodiment of consciousness? Why speak of
incarnate consciousness or consciousness incarnated" (p. 185)? – we could
answer that many thinkers who did *not* consider themselves Cartesians
and who did *not* indulge in any "spiritualist extravagance" and who
seemed to feel they were contesting mind-body-dualism nonetheless *did*
see some sense in speaking this way. Perhaps Spicker may wish to argue
here that those thinkers were in fact unsuspecting Cartesians. In any case,
by the end of this section of his paper, Spicker seems to have defined the
limits of his own use of the "lived-body": "The psychic is nothing but a

H. T. Engelhardt, Jr. and S. F. Spicker (eds.), Evaluation and Explanation in the Biomedical Sciences, 205–208.
All Rights Reserved Copyright © 1975 by D. Reidel Publishing Company, Dordrecht-Holland.

special understanding of the lived-body" (p. 186) and, quoting from
Nietzsche, "... soul is only a word for something about the body." Spicker
clarifies the concluding remarks of his first section by introducing the
term "compositional materialism" (pp. 186–187). For this term he quotes
Professor Joseph Margolis. Margolis, we understand, wants to justify
our speaking of, for example, the Mona Lisa of Leonardo da Vinci as
consisting of matter but not being identical with matter. In this sense
Spicker intends to contribute to those endeavors which attempt "to establish an adequate account of the human condition in terms of a strictly
bodily account of its existence" (p. 183). In this way, Spicker probably
adequately meets the literal demands of a symposium on the interface
of philosophy and medicine.

(2) Spicker introduces the interpretative category of laterality to comprehend the lived-body through its pathology. Pages 191–197 of his paper
bring us to the heart of the matter under the title: "Neuropathology."
These pages are divided into two distinct sections, one on a symptom
complex in neuropathology called "Gerstmann Syndrome," the other on
a related phenomenon, called "Acalculia" (which Spicker defines, referring to the man who coined the term) as "a disturbance of calculating,
produced by a focal lesion of the brain" (p. 194) and its relation to what
Spicker calls the "neglected" Aristotelian category of "position" (p. 196).
Regarding the "Gerstmann Syndrome" not much should be said by me
since I claim no competence in neurology. The issue here is what relevance these neurological findings have for the relation of philosophy and
medicine. On page 193 Spicker indicates that there is a relevance. The
findings are relevant for "philosophers who have a more concrete and
empirical bent and who distrust transcendental inquiries" and who "have
turned to pathological phenomena for clues to the norm of the life of
ánthrōpos" (p. 193). The relevance does not arise, to be sure, from "the
exact location of the lesion in the brain" (p. 193), but it is related to the
earlier alleged fundamental category of "laterality." That this is Spicker's
intent becomes clear when he says that all four elements of the Gerstmann
Syndrome can actually be "understood as a disturbance of laterality –
the directions left and right" (p. 193). The distinction of right *versus* left
sides, he says, is not the product of a "mere linguistic habit" (p. 193);
rather, it "is the condition of the possibility of spatial orientation" (p. 194).
More important, he says, it is the condition of the possibility of "all so-

called 'cognitive' processes, produced by the dominance of one lateral direction over and against the other" (p. 194), e.g., right over left. For Spicker, a location in the brain might be a necessary explication given by neurologists but it is not – as I understand him – sufficient to fully explain a disturbance of laterality. We would like to ask Spicker whether his thesis, that the distinction of right *versus* left is the very condition of the possibility of spatial orientation, is sufficiently analyzed. Is Spicker not a little too hasty here with his conclusions? Does Spicker offer us an adequate account of the presence of the horizontal-spatial meanings of left and right in our bodily perceptual anchorage in the world? The phenomenology of horizontal space, which Spicker has outlined in his interesting paper, recalls the analyses of vertical space which Merleau-Ponty articulates in his *Phenomenology of Perception*. There Merleau-Ponty reflects on the problem of accounting for the presence of vertical-spatial meanings such as "top and bottom" in the perceptual field. Such vertical-spatial directionality can not be accounted for by transcendental reflection in which the unifying activity of a constituting consciousness dissolves the space of the perceptual world into a pure form. Such a form would never allow for the specification of a *preferred* system of coordinates. Spicker is in full agreement with Merleau-Ponty when he rejects advancing a theory of a constituting consciousness in order to comprehend the horizontal-spatial meanings of "left" and "right." But what exactly can we infer from the more or less drastic pathological disruption of (what Spicker calls) "laterality" as to the total spatial orientation of our lived-bodies? Should we not say that there is at the outset an already established directionality, in relation to which the spatial meanings of "top" and "bottom" or of "left" and "right" come into presence?

(3) Spicker writes, referring to an article of Peter Geach, that "the failure to recognize the soliloquistic personal pronoun as merely a grammatical place-holder has led to hopeless searches for the self, for the core of consciousness, and even for a peculiar entity, thus for an only inferred substance or essence" (pp. 185–186). On page 198 Spicker writes, however, that "the totally conceptualized lived-body is not itself a sufficient account of *ánthrōpos*, since 'personhood' too must be accounted for ...." It would certainly be very interesting to know how Spicker would account for 'personhood' in the context of his so-called "compositional materialism" and in view of his endeavors to contribute to an adequate

account of the human condition in terms of a strictly bodily account of that condition. What does Spicker really understand by "a strictly bodily account" of the human condition? We can speak of "*ánthrōpos*" in just the same way in which we speak of a work of art, both consisting of matter but not being identical with matter, Spicker suggests. But it seems to us that the rudimentary means that Margolis, whom Spicker quotes when he deals with the idea of compositional materialism, employs in speaking of *ánthrōpos* do not represent any improvement over Aristotle's distinction of "*archai*" or, for that matter, Kant's distinction between *real* and *logical* existence.

(4) The title of Spicker's paper is "The Lived-Body as Catalytic Agent." We think that Spicker wants to say, with this choice of title, that the lived-body is the basis for any subsequent "sense-conferring," to use an Husserlian expression, by virtue of a "pre-oriented" program built into it. Along the lines of Merleau-Ponty, one would say that because the body, the lived-body, is so "preprogrammed" – by dispositional responses – the *a priori* nature of spatial directionality can be explained. And, that beyond even this, more complex organizations, which in the living organism will have become habitual and hence could serve as the basis for what Husserl calls "active constitution," can be explained as well. If this is correct, it is still not completely clear why Spicker should have chosen the word "catalytic." Does he wish to draw an analogy between the lived-body and a chemical substance which induces alteration without itself suffering alteration?

Duquesne University,
Pittsburgh, Pennsylvania

SECTION VI

THE ROLE OF PHILOSOPHY
IN THE BIOMEDICAL SCIENCES:
CONTRIBUTION OR INTRUSION?

ROUND-TABLE DISCUSSION

H. TRISTRAM ENGELHARDT, JR.

CHAIRMAN'S REMARKS

Recent years have witnessed a birth of interest in philosophical problems
in medicine. This interest spans issues from medical ethics to the philosophy of the science and art of medicine. The result has been the entrance, at times awkward, of the philosopher and philosophy into the
clinical setting. To say the least, there has been suspicion on both sides.
Philosophers have questioned the importance and authenticity of the
philosophical issues in medicine. Physicians have questioned the importance and significance of the contributions to be made by philosophy.
Here and now I hope through you that we can consider the points of
tangency, collision, and contribution between philosophy and medicine.
In particular, I hope that you can say something concerning the possible
effect of philosophy on medicine in the clinical setting, concerning the
ways in which medicine can be enriched, or for that matter distracted
by philosophy. Can philosophy make a positive contribution to medicine
and especially to clinical medicine? And if that is possible, what should
the nature of that contribution be? And as importantly, can medicine
make a contribution to philosophy, can it indicate to philosophy as yet
unexamined or underexamined issues?

Medicine is, in many ways, the science of man in his finitude, the study
of man's life circumscribed by disease and death. Can medicine recall
philosophy to interest in the counsels of finitude, the ways in which the
values of man can thrive or be undermined in the embrace of man's
precarious existence? Can philosophy help medicine to understand better
the nature of the good life, health, which medicine pursues for man? Can
medicine offer new ways for philosophy to understand the nature of explanation and evaluation, the meaning and the significance of the self
and of the body? What can medicine and philosophy teach each other?
It is this rich web of questions about the points of synergy and conflict

H. T. Engelhardt, Jr. and S. F. Spicker (eds.), Evaluation and Explanation in the Biomedical Sciences, 211–234.
All Rights Reserved. Copyright © 1975 by D. Reidel Publishing Company, Dordrecht-Holland.

between philosophy and medicine which constitutes the focus of this round-table discussion.

CHESTER R. BURNS

In speaking of the schism between the humanities and the social sciences, John Higham once declared that "we have too little 'art' in one camp, too little 'science' in the other, and not enough breadth of mind in either."[1] I hope that "breadth of mind" will characterize today's and subsequent attempts to fashion roles for philosophy in the biomedical sciences.

Historically, it is possible to identify three kinds of connections between philosophy and medicine. Every philosopher or physician has been time and space bound. He has received a specific cultural heritage. It is possible, therefore, to examine the thoughts of physicians as they reflect philosophies current at their time and it is possible to examine the thoughts of philosophers for the medical ideas prevailing in their era.[2] But, these connections between philosophy and medicine usually represent a rather passive transmission of cultural legacies, and not the intentional efforts of physicians to be philosophical or of philosophers to be medical.

There are, secondly, examples of physicians who have philosophized, in a general sense, or of philosophers who have utilized medical ideas in developing their speculative viewpoints. These individuals select ideas from each other's intellectual worlds and use them for particular purposes. Philosophers speculate about the meaning of medicine or about health and disease;[3] physicians moralize about appropriate styles of personal and professional conduct.[4] The intent is typically casual, albeit important. As sage and mentor, the doctor must express his wisdom. As a student of wisdom, the philosopher must not ignore man's desires to avoid disease and to live healthy lives.

In the third place, there are examples of physicians and medical scientists who have intended to make significant contributions to philosophy proper and there are philosophers who have wished to affect directly the mainstream of the biomedical sciences.[5] Some of these individuals, particularly physicians, have also wished to shape a realm of knowledge and understanding that they label "philosophy of medicine."[6]

All of these interrelationships between philosophy and medicine are manifested in our contemporary society and remain as potential opportunities for intellectual or social transactions between physicians and philosophers.[7] The inter-personal transactions must be especially encouraged.

It is an unfortunate accident that we scheduled our symposium at the same time as that of the annual meeting of the Texas State Medical Association. This should be avoided in the future because, above all else, personal interactions between health care professionals (including physicians) and professional philosophers are needed. In rubbing shoulders and exchanging ideas, these professionals become aware of each other's problems and concerns. Transprofessional dialogues are facilitated; substantial contributions then become possible.

The interactions also need to occur regularly in non-symposium settings. Ask medical students about the relevance of philosophy as they eagerly discuss their career choices and profound frustrations at 2:00 a.m., seven hours prior to a final exam. Ask the thoughtful physician if he or she is reflectively assessing values when making a decision about the best of two drugs or the best of two surgical procedures or the best of two ways of dealing with emotional difficulties. Ask a philosopher if his ideas of medicine change when a parent or spouse or child is hospitalized. Ask a graduate student in philosophy to reflect on observations made on a Saturday night in a hospital's emergency room.

Other educational opportunities can be provided that enable health professional students to meet their crises, physicians to resolve their dilemmas, and philosophers to enlarge their perspectives. These academic programs are developing within a "humanities" context.

Philosophy is viewed frequently as one of the humanistic disciplines. But, when the words – humanitarianism, humanities, and humanism – are tossed about casually, confusion arises for both physicians and philosophers.

All three have something to do with a devotion to human affairs, with a regard for the interests of humanity, and with a desire to advance the welfare of mankind. Humanitarianism connotes the attitudinal dimension of being human – compassion, empathy, courtesy, kindness – the art of being considerate. The humanities have something to do with the cognitive dimension of man, with learning or literature that express human activities distinctively – the art of being reflective, conversant and literate. Humanism, in addition to the specific mode of education prominent in the early modern era, refers to a system or style of thought or action concerned exclusively with human affairs; that is, a philosophy of man including a set of values that reflects a consciously determined life perspective leading to a consistent and satisfying life style – the art of being a competent human.[8]

Those who wish to relate philosophy to medicine must attend to all three dimensions. They should demonstrate that the art of being reflective, conversant and literate cannot be taught without relating it to the art of being compassionate and the art of being competent. Quacks cannot make these demonstrations.

There are quacks, men and women who understand neither philosophy nor medicine, but who pretend to solve "bio-ethical" or "socio-linguistic" problems. These medico-philosophical intruders need exposing.

Other well-intentioned individuals have distorted images of the aims of philosophy. I am distressed equally by physicians who claim that medical ethics never change and by philosophers who claim that "medical" ethics do not exist. I am especially puzzled by those – be they physicians or philosophers – who view most doctors as illiterate, debased technicians or those who view philosophers as fascinating ne'er-do-wells. These misconceptions need correcting.

Far more than exposé or corrections, though, we need vision, courage, tolerance, and forbearance.

Those who wish to bridge philosophy and medicine in their research and teaching must convince skeptical professions (and publics) of their right to pursue their goals. Accordingly, they have a challenge more difficult than that of the orthodox physician or that of the traditional philosopher. But, as Professor Temkin said of physicians-historians of medicine, "it is hoped that in the end, they will deserve the respect of both."[9]

University of Texas Medical Branch,
Galveston, Texas

NOTES

[1] John Higham, "The Schism in American Scholarship," *American Historical Review* **72** (1966), 20.

[2] An example of the former: Owsei Temkin, "The Philosophical Background of Magendie's Physiology," *Bulletin of the History of Medicine* **20** (1946), 10–35; an example of the latter: Lester King, "Plato's Concept of Medicine," *Journal of the History of Medicine and Allied Sciences* **9** (1954), 38–48.

[3] Morris R. Cohen, *The Meaning of Human History* (La Salle, Illinois: Open Court, 1947), pp. 199–213.

[4] *A Way of Life and Selected Writings of Sir William Osler* (New York: Dover, 1951).

[5] For an example of the former see the English translation of Galen's "Institutio Logica" prepared by John S. Kieffer (Baltimore: Johns Hopkins Press, 1964); for an example of the latter, see H. Tristram Engelhardt, Jr., "Explanatory Models in Medicine: Facts, Theories, and Values," *Texas Reports on Biology and Medicine* **32** (Spring 1974), 225–39.

[6] Claude Bernard, *An Introduction to the Study of Experimental Medicine* (New York: Dover, 1957); also see Lester King, "Medical Philosophy, 1836–1844," in *Medicine, Science, and Culture. Historical Essays in Honor of Owsei Temkin*, ed. by Lloyd G. Stevenson and Robert P. Multhauf (Baltimore: Johns Hopkins Press, 1968), pp. 143–159.

[7] George E. Arrington, Jr., "The Medical Philosopher – His Role in Contemporary Medicine," *The New Physician* **12** (1963), 159–63; K. Danner Clouser, "Philosophy and Medicine, the Clinical Management of a Mixed Marriage," *Proceedings of the First Session of the Institute on Human Values in Medicine* (Philadelphia: Society for Health and Human Values, 1972), pp. 47–80.

[8] For some of these ideas, I am indebted to Edmund Pellegrino.

[9] Owsei Temkin, "Academic Future of the History of Medicine," *Journal of Medical Education* **39** (1964), 792–793; also see Temkin's essay, "On the Interrelationship of the History and the Philosophy of Medicine," *Bulletin of the History of Medicine* **30** (1956), 241–251.

JEROME SHAFFER

Every celebration should have its Jeremiah crying "Woe to those that are at ease in Zion." I propose to carry out that role here. I wish to raise some doubts about the validity and value of the Philosophy of Medicine.

I note that this conference defines its concern not as the Philosophy of Medicine but as "the interface of philosophy and medicine." [1] This indicates a salutary caution on the part of its conveners. Perhaps it was just courtesy, since "Philosophy of Medicine" might have been thought to put the nonphilosophers at a disadvantage, whereas "Medical Aspects of Philosophy" might have unduly depressed the philosophers. Still, to refer to "the interface" of philosophy and medicine is to flee to metaphor, presumably drawn from the hard sciences, perhaps to give an impression of rigor and precision. The expression, "interface," does suggest that there are some problems common to philosophy and medicine,[2] but it is not at all clear to me that there are any such common problems. I am inclined to think that there are medical problems and there are philosophical problems, with no overlap or borderline area between them, no field which could be called medico-philosophy or philosopho-medicine on the analogy with bio-chemistry or astro-physics.

Still, there could be a field of philosophy to be called Philosophy of Medicine, on the analogy of such fields as Philosophy of Art, Philosophy of Science, Philosophy of Law, Philosophy of Religion, etc. To that hallowed triumvirate, Truth, Beauty, Goodnèss, the man on the street would add Health – he might even add the ontological claim, that when you have your health, you have just about everything. The philosopher asks, What is Health?, and comes to Galveston for the answer.

I am inclined to think that the use of the heading, Philosophy of Medicine, is misguided because as we shall see, it turns out that most of the problems alleged to fall under that heading are either already properly handled elsewhere or not properly philosophical at all.

I do believe that there is an important field of philosophy called Philosophy of Science and an increasingly important branch of it entitled the Philosophy of Biology. The philosophical problems concerning the definition and nature of life, the kinds of laws governing living things, the kinds of analyses and explanations appropriate to living things, the concepts of cell, gene, species, instinct, race, etc., and their places in our total conceptual scheme, and so on, will naturally fall under the heading of Philosophy of Biology. How about the concept of what for an organism is normal and abnormal, well-formed and malformed, healthy and unhealthy, non-pathological and pathological? Here I agree with Professor Toulmin that these are part and parcel of the theory and laws of biology and I disagree with Professor Wartofsky and Professor Engelhardt that these notions are ethical, aesthetic or even social in nature. I see here an undesirable tendency to extend Szasz's thesis concerning mental illness and proclaim what might be called "the myth of physical illness," the view that disease and diseases involve in their very *definition* moral and social responses to physical structure and functioning. On this view health, like beauty, is in the eye of the beholder. This seems to me a mistaken view. House-maid's knee is a genuine malady, even if it may be caused by capitalism and male chauvinism and its cure may involve the radical restructuring of society. After all, concepts of health, disease, abnormality, malformation, stunted growth, etc., apply quite uncontroversially to plants and animals as well as to people. It is an historical accident that "doctor" applies primarily to someone who treats the maladies of *people*. If we valued plant and animal life as much as we do human life, the health profession would include gardeners and conservationists as well.

What more is there to medicine than the science of biology? There is the *practice* of medicine, the activity by which we attempt to intervene in the world and so change it as to realize the goal, purpose or intention of medicine, namely, preventing, alleviating or eliminating *undesired* abnormalities, malformations, pathological or unhealthy states. This means that another putative source of subject-matter for Philosophy of Medicine is the study of the political, social and ethical issues that arise in the process of practicing medicine. To decide these issues, the practitioner may have to commit himself on difficult questions concerning who should live and under what conditions, what is just, what is legal, the comparative value of different lives, the weights of various duties to himself, his family, his profession, his patients, his society, and the world and the future, what actions will maximize happiness or minimize pain, what is the greater or lesser of a number of possible evils, what the probabilities of various outcomes are, what religious or metaphysical speculations are true or false, and what to do when answers to the above questions yield conflicting courses of conduct. One might think that this is where philosophy comes in. I think we would all agree that it is not in the province of philosophy to provide *answers* to such questions of conduct. What can philosophy do here? Well, we can look to Moral Philosophy for assistance. It can provide some clarification of the issues, the elucidation of hidden assumptions, an examination of the kinds of arguments used, a tracing out of the relations between issues and of the relations between the concepts involved in those issues, a separating out of the more factual and the more evaluative elements in a discussion, etc. All of this is very important and valuable, but I believe it is important and valuable *in and for itself alone* rather than for its help on reaching solutions. On none of the questions concerning the political, social and ethical issues is it likely that philosophers will agree. If anything, they will tend to add to the difficulties and make it harder to come to any decision whatsoever. Consider abortion, for example. Surely the Supreme Court was correct in pointing out that "those trained in... medicine, philosophy and theology are unable to arrive at any consensus." Here I agree with Professor MacIntyre's conclusion that "the medical profession ought not therefore

to look for solutions [to moral problems] to philosophical theorizing."
At best, philosophy can offer its own special perspective, its own peculiar
sort of understanding. The philosophic vision is important in itself even
if it does not yield or advance particular solutions to practical problems.
When philosophy has done its work, the great moral and social issues
will remain. In the end philosophy will bake no bread here.

So far, then, I have argued that the work to be done by a putative
specialist in Philosophy of Medicine can best be done by those skilled
in Philosophy of Science and Moral Philosophy. I am sure many qualifications of my thesis are in order, but two strike me as particularly important. First, as Professor Toulmin restricted his conclusions to *somatic*
medicine, so would I also. The concepts of mental and emotional health
and disorders and the allied concepts are fundamentally problematical.
Here it is Psychology which must be added to Biology, and the Philosophy
of Psychology or the Philosophy of Mind would be the relevant fields of
Philosophy in which the problems arise.

Second, so far as the papers of Professor Zaner and Professor Spicker
are concerned, I am not clear how to assess the contributions of Phenomenology. Many of their remarks seem to be in the mainstream of
metaphysics and mind-body speculation with particular importance being
given to certain findings of neuro-physiology and psychology. What
might be left, exploration of the subjective, what-it-feels-like content of
concepts of health and illness, the doctor-patient relationship, etc., might
be a part of a more developed psychology.

To summarize, then, if we do our Philosophy of Science, Moral Philosophy and Philosophy of Mind properly (or Philosophy of the Body,
as Professor Spicker would have it), there is nothing left for the Philosophy of Medicine to do. On the other hand, nothing I have said here is
intended to minimize the great importance of the issues we have been
discussing here. I can think of nothing *more* important in philosophy and
would even sanction the use of the incorrect title, Philosophy of Medicine, if further discussion of these matters were thereby encouraged.

University of Connecticut, Storrs, Connecticut

NOTES

[1] It should be noted that the full title of this Symposium was "First Trans-Disciplinary
Symposium on the Interface of Philosophy and Medicine."

² In the discussion, it was mentioned that "interface" often is used to refer to two proximate boundaries with little exchange across them, as happens with water and oil. The
premise of this conference is that there is or should be *no* interface, in that sense, between
philosophy and medicine.

KENNETH VAUX

On behalf of the several journeymen from Jerusalem who have wandered
into Athens these days, let me thank you. Although your deliberations
expressing the limitations of philosophy have created an awesome statue
to the unknown God, I will desist from proclamation, unlike that other
Jew who happened into the philosopher's garden. Rather, I express gratitude for the precision of thought and the candor of these meetings. Even
though we haven't seen the smoke of JAHWEH to lead us, or at least etch
a few guidelines on the rocks, we've done damn good iconoclasm, which
is the *via negativa*, the first step. At least we've opened the moral abyss.

If I'm hearing those who have addressed us correctly (Wartofsky,
Burns, MacIntyre, Engelhardt, Toulmin, King, Spicker, Pellegrino,
Zaner, *et al.*) these days, I sense the distance between Athens and Jerusalem is getting short. This was accomplished surely as much by meeting
on this lovely Island as by Henry Kissinger's jet plane.

This observer senses that MacIntyre's counsel that modern philosophy
move beyond those periods of paralyzing analysis and dilettantism to
concerns of genuine significance may, in fact, be occurring. It is my hope,
as a theologian, that philosophers will not only continue to demand of
all of us clarity of language and meaning, but will grapple with great
questions. I hope that some few will even recover that ancient foolishness
to become theologians and moralists. On the practical level, I am in-
trigued by Gorovitz's willingness to see the philosopher as therapist, cognitive of course. We need philosophers' involvement in the clinical processes of moral analysis, decision-making, and advocacy.

Philosophy has a contributing, indeed an essential, role to play in bio-
medicine and clinical praxis. Asking questions and sharpening analysis
is helpful, but not sufficient. Philosophy must recover practical wisdom,
Phronesis in Aristotle's sense. This Socratic mode of dialectic, where the
true and the good are seen as self-evident when the layers of deception
are carved away, will best serve us. The cynic and skeptic modes of asking
questions for question's sake, however refreshing in times like these, will
not serve the resolution of the kinds of problems we face in bio-medicine.

I would like to propose a way that *Sophia* and *Phronesis*, those wisdoms essential to understanding and sound action, can be brought to bear on these profound areas of human experience. Phronesis is defined by Aristotle as "a rational faculty exercised for the attainment of truth in things that are humanly good and bad."[1] I believe this proposal fits the modest capacity of philosophy that most good philosophers acknowledge.

I wish to speak of three instruments of human understandings: Science, Philosophy, and Theology. Into each I pour much more meaning than the words denote. *Science* I would define as the empirical activity which is the continual testing of hypotheses in the realm of the structure and process of nature. *Philosophy* is the formal process of making connections. *Theology* is that range of perceptions that sense transcendence and posit therefrom ultimate meanings and values. Here we must include the arts and literatures in addition to religion.

Because of the character of the ethical problems in bio-medicine, philosophy is needed to play a mediative role between the empirical and the normative processes in human experience. The analytical, interrogative, logical disciplines of philosophy can effect the conjunction of the activities of scientific medicine translated into care of persons, with the insights of the true and the good perceived in hope, imagination, and faith.

Prophetic (faith generated) values which elicit anger at what is, because of what ought to be, and the proleptic (hope generated) values which signal what could be – they alone can speak to the deep cleavages in human existence opened up to us in advancing bio-medicine. Philosophy is needed to transpose these insights into concrete public and personal decision-making.

The powers of bio-medicine – generating, regenerating, refashioning human beings, along with the accompanying drama of suffering, death, and metaphysical guilt – call from our depths apperceptions of transcending meaning and goodness. Morality is the translation of the impulses of faith, hope, and love into *caring action*.

Philosophy can best perform its task by four activities: questioning assumptions, discerning motives (here, I fear, cognitive analysis will not suffice; hopefully the Socratic therapeutic graces of honesty and confrontation have survived your Ph.D.), fracturing idols, and developing schemata of options. Philosophy can bring the sobriety and humility essential to goodness by exposing the provisional character of hypotheses (e.g., disease categories), and the contingent character of clinical decisions. In short, good philosophy can fracture life open to mystery, yielding the abyss or perhaps the vision of God and the good.

Ideally the scientist, philosopher, and theologian should live in every person, certainly any professional who intends to serve in the arena of human life, suffering, and death. This kind of personally integrated person can best translate the normative into the bio-medical. Until that day when good men make good decisions (despite reams of medical testimony before Mondale and Kennedy Commissions, that day is not yet here), the intrusion of philosophy into bio-medicine will be as necessary for re-conception as intercourse used to be for conception.

Let me conclude with some practical proposals. We need medical clinicians who are philosophers and bio-ethicists. There are a few around, most too busy taking care of folk and their house staffs to be here. There are some young doctors with us who will be such. We need more fellow-ships, centers, foundations to train such. Thanks to NEH, the Kennedys, The Society for Health and Human Values, etc., we have a start. We need dialogue between the physician, the pastor, the philosopher – yes, even the lawyer, because he picks up the pieces when moral consultation and advocacy have been unavailable. What is happening in medical schools is not nearly as important as what must happen in clinics, in hospitals, with house staff and with parents and families. This is where medical education and ethical formation take place. Some of us work to help the minister to be part of the answer rather than the problem, others work with the physicians, others with counselors and social workers. I wish there were a healing profession the philosopher could assume in order to take his place on the health team. The *ombudsman* may be an option, but that will require philosophy chastened with passion. I con-clude with Leo Strauss:

> "It is the philosopher's business to transmit
> wisdom which begins in wonder.
> It is the theologian's business to transmit wisdom which
> begins in the fear of the Lord."[2]

Institute of Religion and Human Development, Houston, Texas

NOTES

[1] Aristotle, *Nicomachean Ethics*, 6.5.1140b. *The Ethics of Aristotle*, trans. by J. A. K. Thomson (London: Penguin Books, 1953), p. 177.

[2] Leo Strauss, "Jerusalem and Athens: Some Introductory Reflections," *Commentary* (June, 1967), pp. 45–57.

MARX W. WARTOFSKY

Speaking as an "Athenian Jew," I would like to reflect on the question of integration which Dr. Vaux proposes, and also to protest the alleged schizophrenia between the "Jew" and the "Athenian" (or between Dionysos and Apollo, or between the "practical" and the "theoretical," or the "ethical" and the "rational"). It seems to me that we wrongly conceive the relation between philosophy and medicine if we see philosophy as some sort of "Apollonian" or bloodless "Reason," standing by dispassionately and doing its cognitive therapy, while the surgeon dirties his hands and bloodies his patient. It is my view that this dichotomy must be overcome, if philosophy of medicine is to become more than a formal game to keep philosophers from getting bored or to provide physicians with ornamental modes of after-hours "thinking."

Being an "Athenian Jew," and concerned therefore with both reason and praxis, with both thought and the passions, I think what is needed is a science of the passions, a science of practice, *and* a passionate science (indeed, a *com*passionate science). Spinoza, in the seventeenth century, attempted such a science of the passions, from which we have yet much to learn. The problem with which Spinoza grappled was how human feelings were to be put in touch with reason – not "under the control" of reason – and how reason itself was to become capable of subserving human practical ends, human action; in short, how reason was to be integrated in the whole person, so that the life activity of human beings would be freed of pathologies of thought and action. The model here is one in which the most theoretical philosophy is undertaken in the context of the most practical human ends. It is a model which one could profitably emulate in considering the relations of medicine and philosophy.

In considering the relation between these two fields, the discussion here has been provocatively shot through with sexual metaphors: the "intrusion" upon or "penetration" of medicine by philosophy; the "marriage" of the two; the "fructification" of one by the other. In such suggestive contexts, even the term "interface" begins to connote coital position, rather than its more innocuous senses (borrowed from physical chemistry, as we have been told.) To pursue the metaphor (perversely), I should note that very much depends on what we take the relationship to be: whether it is one of rape or seduction of one field by the other; whether it is a proper marriage of mutuality, and whether it is consummated and bears fruit, or whether it remains one of occasional joint tittilation by both parties. In both fields, as in many of the disciplines which have somehow been jammed together willy-nilly, there is a fear of losing autonomy and losing the distinctiveness of one's subject matter, resulting in some fluffy cracker-barrel wisdom – with little relation to the hard facts and rigorous criteria of the original and independent disciplines. This is a real fear. But let's separate it from the equally real possibilities of success which may be achieved. In keeping with this theme, I would like to speak to several of the issues which were raised here.

I think that Dr. Vaux's suggestion that real wisdom, *Phronesis*, is integrated with practice is an instructive and useful beginning. *Phronesis* is best understood not merely as a contemplative mode of wisdom, but rather as an active one: the kind of active wisdom which it is incumbent upon philosophy – no less than medicine – to seek. In this sense, the view that philosophy should perform a merely contemplative or analytic or auxiliary therapy in the contexts of medical practice is simply wrong. It presents the wrong model, and the wrong goal. Now, I for one don't want philosophers to do surgery or diagnose illnesses. But I do want philosophers and physicians alike to reflect actively upon the concrete praxis and theory of medicine. Nor is this a philosopher's holiday – an excuse to go "slumming" in the wards or operating rooms. I am selfish enough about philosophy and its proper perquisites, its demands for rational rigor, conceptual clarity, self-conscious criticism. But I think that in order for philosophy to continue to count (or to *begin* to count once again), indeed, for it to thrive and prosper, it has to be destroyed in its present modes, it has to be "overcome," or "*aufgehoben*," as we dialectical types like to say. Philosophy must overcome its divorce from practice. To do so, philosophy has to seek its resources where it has always sought them, and where it was originally generated, namely, *outside* of philosophy. Philosophy becomes effete, it becomes formal, it becomes sterile, whenever it seeks to have some pretended autonomy devolve upon itself, to acquire a separateness and discreteness which leaves it with no practical content. It then becomes sheer formalism, or mere virtuosity.[1]

Now, with respect to the question of whether the philosopher is, or is not, to give solutions to moral problems, or whether he or she is merely to theorize about them, I would consider MacIntyre's critique of philosophy, and of its inability to give answers and solutions to moral problems *not* as a critique of some *essential* limitation or incapacity of some timeless enterprise called "Philosophy." Rather, I take him to be making an historical critique of the *contemporary* incapacities of philosophy to answer such questions. On this view, one would say that, while the philosopher is not now in a position to provide solutions, he ought to be. Thus, I would agree with Gorovitz that philosophy should undertake to provide moral solutions and answers to practical moral questions. But it should do so, not as an outside agent, from some independent Mt. Sinai, presenting ready solutions, a new Decalogue, to the lawless masses who suffer for want of the beneficence of philosophical grace. Rather – and I think Gorovitz would agree – philosophy becomes relevant here by virtue of its immersion in those practical contexts which generate *Phronesis*. Without such immersion, philosophy remains not only unbaptized, but unsalvageable and unredeemable.

But enough of Old and New Testament imagery. Another question looms: What will "Philosophy of Medicine" be like? Will it be like its presently acceptable correlates, e.g., Philosophy of Physics, or Philosophy of Biology? I.e., will it become the latest branch (or twig) of the "Philosophy of Science" tree?

One way of formulating the problem is as follows. Even in its "applied" contexts (e.g., in philosophy of science, or education, or religion, or art), philosophy proper is concerned with theory: with the critique of theory, with theory-construction, with the logic of theoretical or critical discourse. It removes itself from the practical contexts of application which are the proper domain of these various enterprises and therefore do not fall within the purview of "philosophy proper." The paradigm of what constitutes a proper domain for philosophy has been theoretical physics, a "real science," for there, the matter of practical or anthropomorphic concerns has been eliminated. In physics we are not so caught up in human caring as to distort our view. This leaves us dispassionate enough for the kind of critical reflection which philosophy demands. All other sciences *become* scientific, on this view, when they approach the condition of mathematical physics.

The paradigm of mathematical physics was established in the seventeenth century. The paradigm of *really* "scientific" biology appeared only in the twentieth century, when biology began to emulate the paradigm of mathematical physics. *Really* "scientific" behavioral science develops, on this view, again only when it approaches this condition; and the latest of the additions to the Philosophy of Science *proper*, the Philosophy of the Social Sciences, also achieves respectability when it too becomes the latest and most problematic branch of the physical science model. Since medicine is far removed from such a paradigm (like History, about which one also asks: "Is it a science, or is it an art?"), it is not yet respectably a subject matter for such scientific philosophy. At best, it may be left to the unscientific, or anti-scientific and feelingful rhetoric of the "soft" philosophers: the poetizers of the "life-world," the phenomenologists, muddy ethicists, and the existential types.

It is only recently that this positivist paradigm of science itself has come under serious question by the philosophers of science themselves. This philosophy of physics, we have learned, has very little to do with the way physicists actually work, think, discover, explain, etc. Whether medicine has to meet some intrinsic criteria of "scientificness," in order to become proper grist for the philosophical mills is, therefore, not established *a priori*. Indeed, doctors will have very little to learn from philosophers, and philosophers from doctors, if we attempt to impose a particular historically-evolved conception of philosophy of science upon medicine – namely, that of the positivists. The origins and aims of the positivist school are, in themselves, an important subject for historical and sociological, as well as philosophical analysis. The specific *Problematik* which positivism was developed to cope with was the task of overcoming a particular form of obscurantism which impeded scientific thought. It was, therefore, in its time, a radical break with a particular mode of obscurantism, clericalism and anti-humanism. It is the weakness of *a*historical appropriations of the positivist insights which have led to a new dogmatism and obscurantism, which, to my mind, stands in the way of new philosophical advances, in dealing not only with the traditional sciences, but with medical theory and practice as well. If the philosophically-prone doctors give up muddle-headed cracker-barrel "wisdom" for an equally irrelevant pretense at "logical-philosophical rigor," borrowed from an historically anomalous view of what "real science" is about, then what will result is not a new "Philosophy of Medicine," but only the latest in a series of fads, whose relevance to live medical practice and theory will be negligible. Just as those doctors who do the kind of "ethics" which Dr. Burns was talking about are charlatans and quacks, so too will the new "positivist" philosophers of medicine be charlatans and quacks. Whether the charlatanry and quackery are characterized by "cracker-barrel wisdom" or by the adoption of the positivist jargon of the philosophy of science relevant forty years ago, is incidental. It will still be quackery, whatever its mode.

Finally, I would say, in answer to Shaffer's critique, that, although his critique is thoughtful and warns against giving up the hard medical facts for a mess of philosophical pottage, he needs to be clearer about what he is objecting to. I don't think either Engelhardt's or my proposals were of the sort that Shaffer makes them out to be. Now housemaid's knee is undoubtedly a disease of the knee, a somatic illness, a prepatellar bursitis. There is no argument about that, nor about the fact that it demands somatic treatment. The point of my paper was not to deny that. Rather, it was to insist: (1) that one *also* needs to take into account *medically* such contexts as the etiology of the disease, its social and historical characterization,[2] and (2) that such considerations are as much a part of the *medical* concern with that disease as is the concern with its narrower contexts somatically. My argument was that the context for the treatment of a disease is never simply the one in which a particular patient in a given case in a given hospital or office setting is being "treated" (though it is always *also* that), but that the disease *as a disease* is an entity which has a spatio-temporal existence beyond the given particular instance, and that therefore the treatment of the disease in a given instance demands as much of a consideration of the social, historical, and etiological contexts as it does of the more narrowly somatic and therapeutic contexts. In fact, in the absence of such considerations, there would be no learning from experience in medicine, and there would be no medical theory. Medicine would forever remain an *ad hoc* craft or skill (if even that, since no principles could ever be formulated or transmitted). Thus,

I do not mean to counterpose the practical treatment of a given patient to the larger contexts, but rather try to place them in relation to each other. In doing so, we come to see that they are not incommensurable.

I would argue, in effect, that just as we have to widen our vision in philosophy, and get involved in the nitty-gritty contexts of medical practice, so, too, the physician has to widen his or her vision of medical practice – not simply as an individual, but in the socialization of the profession. Let us take the philosophy students down to the emergency wards, let us take them on rounds, let us get the philosophy professors working in teams with the doctors (and with sociologists and others) in the hospital setting, and in the medical schools. Let us get the medical students to take humanities courses not as ornamental frills for the sake of their "general culture," [3] but rather as contexts for a humanistic understanding of medicine itself, its history, its social contexts, its methodological issues, its epistemological and ontological dimensions. Let the doctor be as specialized as is required for successful understanding and treatment of disease; but let him also understand himself as an historical human agent in a concrete social context. And let him learn to become critical of his profession in a rational and socially responsible way.

I look forward to a much more active (and even problematic) development of relations between philosophy and medicine. I would hope for the development of concepts of the human which practically affect medical modes of diagnosis, of therapy, and the institutional forms in which health care is practiced. For example, I would agree with Dr. Spicker that we need a concept of the body and the person which extends beyond the narrower inherited traditional views of somatic medicine. For although somatic medicine is a necessary abstraction of the human subject and of human disease, and although it continues to require the specialization and division of labor without which medical advances could not be made, and medical treatment could not succeed, that very specialization and division of labor need to be reintegrated now, at a higher level. To foster this change, I would propose that the *interface* here become a much more active matter of *intercourse* between the two fields, that a fructifying intercourse be found, and that appropriate institutional structures be developed.

In keeping with the "theological" mood of some of the previous discussion, I should like to close with a passage from "Scripture," in this case, from Karl Marx's Eleventh Thesis on Feuerbach: "The Philosophers have only *interpreted* the world in various ways; the point is, to change it."[4] I believe that we philosophers should get into medical contexts in practical ways also for our *own* sakes, selfishly, so to speak; for I believe that philosophy has to get beyond philosophy in order to remain philosophical.

Boston University,
Boston, Massachusetts

NOTES

[1] That formalism may take many forms. One of the forms it has taken is the irrelevant attempt to be relevant in applied philosophy, in contexts such as the philosophy of science, the philosophy of religion, etc. I am all for applied philosophy, but one of the characteristic modes of applied philosophy is to keep the philosophical tools clean and only to dip now and then into the muddy waters of this or that practical enterprise and look for demonstrative examples in order to teach the students how to understand philosophical concepts.

[2] See Dubos, "The Diseases of Civilization" and Rosen, "Medicine as a Function of Society," in *Mainstreams of Medicine*, ed. by Lester S. King (Austin and London: University of Texas Press, 1971).

[3] A medical acquaintance once remarked to me, concerning the humanities, that he thought they were very important – he took an English course once at Harvard, and has enjoyed reading literature ever since. Obviously, this is *not* what I am talking about.

[4] Karl Marx, "Theses on Feuerbach," in *Writings of the Young Marx on Philosophy and Society*, ed. by L. Easton and K. Guddat (New York: Doubleday, 1967), p. 402. *Marx Engels Werke*, Bd. 3 (Berlin [DDR]: Dietz Verlag, 1969), p. 535.

EDMUND D. PELLEGRINO

As the only physician on this morning's panel without the redemptive grace of an additional earned degree, I must approach the philosophic questions before us with diffidence. I am chastened on the one hand by MacIntyre's reserve and Shaffer's Jeremiad, while awed on the other hand by Wartofsky's and Vaux's enthusiasm for a dialogue between "homo philosophicus" and "homo medicus." Being a member of the latter species, I must diligently avoid becoming one of Robert Burton's Philosophasters[1] as I respond to the comments of my learned philosophical colleagues, or at least gloss the texts of their remarks.

The questions before us are whether or not philosophy has a legitimate contribution to make to medicine and whether a philosophy of medicine is possible in so derivative a science and art as medicine. I believe both questions should be answered in the affirmative. Indeed, I will go further and state my conviction that a deep and continuing engagement between medicine and philosophy is necessary for the intellectual health and the social and cultural mission of each discipline.

As some of the discussions in this conference have already shown, there is danger in such an engagement. Medicine must not become poor philosophy, or philosophy poor medicine. That peril has too often befallen these venerable partners over the centuries of their relationships, as first one and then the other was culturally in the ascendant. If the engagement is to be mutually beneficial, each discipline must possess the intellectual strength to remain independent, yet be willing to engage the other on substantive issues.

For reasons I have detailed elsewhere, I believe that medicine and philosophy today are approaching another period of conjunction – one which promises to be as fruitful as any they have yet experienced. Philosophy has passed through its eighteenth and nineteenth century overindulgence in speculation and its twentieth century overindulgence in analysis and introspection. It seems ready once again to look at first order questions and at the meaning of man and his existence.[2]

Medicine, in its turn, has freed itself from the subservience it so long suffered to philosophy and overspeculation. It has been enormously enriched by its close identification with the physical and biological sciences, from which it has drawn a vast reservoir of verifiable phenomena about man. Medicine has overindulged somewhat in positivism and reductionism, but it is awakening to a corresponding need for synthesis and for examining its own ends and purposes.

Only once before – at the apex of Greek Culture – have the potentialities of interaction been as great. The present-day opportunity could exceed even that enjoyed by Attic philosophy and Hippocratic medicine. These two strong intellectual endeavors can now meet each other without fear of capitulation, and out of that engagement contribute both practically and theoretically to human knowledge – perhaps even to the elaboration of a new Paidea.

There are clearly two ways in which philosophy can engage medicine – as philosophy *in* medicine and as philosophy *of* medicine. From what my philosopher colleagues have said in this conference, there is little

dispute that philosophy in medicine is a possibility – after all that is
what they have been offering us for the last two days, though with varying degrees of reserve and gingerliness. Philosophy *of* medicine is a more
troublesome and a more problematic concept. Indeed, its possibility is
categorically denied by Professor Shaffer and fastidiously avoided by
Professor MacIntyre. I would like to consider both modes of engagement
briefly.

Before doing so, I want to applaud the "economy of pretension"[3]
practiced by both Shaffer and MacIntyre with reference to what their
discipline can contribute to medicine. Their reservations are an admirable
antidote to the hubris which so easily afflicts both our disciplines. I do
hope they are not protesting too much, however! One way to avoid the
troublesome issues would, of course, be to deny them. But, this seems
unlikely. Our philosophers have, after all, taken the trouble to come to
the conference; they have sincerely looked at the issues. They have, in
fact, illuminated those issues, thereby demonstrating that the philosophic point of view and the analytic mode can at least rigorously define
for us what philosophy *cannot do* in medicine.

Their point is an important one. We must not expect the philosopher
to be the resident intellect or the resident ethicist, ever available to relieve
physicians of the responsibility for thought or to provide the "right"
formulae for ethical actions. It will not distress either Shaffer or Mac-
Intyre to learn that some of us never had that expectation. What we do
expect is the exercise of the philosopher's critical intelligence and his
special way of questioning what he sees in medicine. This is what I mean
by the first of the two medical uses of philosophy – philosophy *in* medicine – the exercise of the philosophic intellect on the data, the reasoning
and the value choices which constitute modern medicine.

Medicine today has the unprecedented capacity to alter profoundly
the lives and the behavior of individuals and society. Uncritical application of the capabilities of medicine is an ever-present danger for society.
There is a need to subject medicine's axioms and assumptions about its
purposes and values to rigorous inspection. Otherwise the authority of
the clinician at the bedside and in the operating theater may be transferred *in toto* to the realm of values – individual and social.

Philosophers do have a social role to play as "delegated" intellects, as
thinking advocates for society. Philosophy *in* medicine can uncover the

pre-logical assumptions upon which medical actions may depend; it is needed to expose some of the limitations of the scientific method, to look at the same questions of causality, classification, logic and epistemology which medicine shares with the other sciences.

These same questions are raised by the philosophers of science, but they are vastly more complicated in the medical context. Medicine deals with subjects in whom the dimensions of purpose, consciousness, reflection and self-determination are predominant. If man is the most complex of beings, the philosopher should be interested in his experience with health and its lack, since this is so profound and universal an experience. Even those philosophers who look upon man as only a physico-chemical system must agree that he presents unique levels of complexity in his behavior.

Philosophy *in* medicine, therefore, includes every one of the classical branches of philosophy, bringing the philosopher's way of looking at reality into closer contact with the whole array of data and experiences we call medicine. Each of the philosophers in this conference has done this in some degree or other, and there is little need to justify this medical use of philosophy further.

What is much more problematic is the question of the possibility of a philosophy *of* medicine. Professor Shaffer flatly rejects this possibility. He would reduce the philosophy of medicine to the philosophy of science, or the philosophy of biology and/or psychology. He denies that a philosophy of medicine can exist in the same sense that a philosophy of history, science or art can exist. His fundamental point is that medicine itself does not exist as a separate discipline, but it is merely the sum of its component sciences. Man then must be the sum of his parts, not something more, or something unique, since man is the subject of medicine, though from the special point of view of health and illness.

Professor Shaffer, while denying the possibility of a philosophy *of* medicine has, of course, philosophized *about* medicine – what it is, and how it relates to the other sciences, as well as making assumptions about the nature of man. He has, in fact, exhibited some elements of a philosophy of medicine – even in the process of denying its possibility. But even his *via negativa* is useful and essential to the question of a philosophy of medicine – for it illustrates one of the ways in which we may regard medicine ontologically.

There is an alternate view which requires close examination. This view holds that a philosophy of medicine is possible, and that it is more than the numerical total of the philosophy of the individual sciences which make up medicine. Our acceptance of this alternative turns critically upon our metaphysics of man. Medicine is today the focal point of our knowledge of man, uniting information from every science and even the social sciences and the humanities. It unites those disciplines – insofar as they deal with man in the experience of health and illness – in a special way – called the clinical mode. This way of thinking and organizing our knowledge of man is more than a summation of knowledge from the physiological and psychological sciences. Hence medicine has a conceptual framework not contained in the other sciences, and medicine and its subject – man – are more than the summation of the component parts of which they are constituted.

This thesis admittedly must be demonstrated. I have not the time to do so here; I set it before you for further examination lest Shaffer's and MacIntyre's justifiable, but not conclusive, reservations about a philosophy of medicine may receive some balance. Some rudiments of the conceptual framework for a philosophy of medicine are contained in Scott Buchanan's *Doctrine of Signatures*,[4] where the structure of medicine is examined with classical metaphysical tools and Foucault's *Birth of the Clinic*,[5] where medical thinking and perception are subjected to a historical analysis.

A philosophy of medicine is possible and essential because medicine unites physiological and psychological, as well as chemical and biological, concepts and modes of explanation. Unless we wish to reduce all of these sciences to one – and Shaffer has not suggested that – then we need some synthesizing discipline to mesh one mode of explanation with another. This is more than the exercise of philosophic modes of thought *in* medicine. It is rather a philosophic – analytic, logical or epistemological – examination of what medicine is and does.

A philosophy of medicine is needed to deal with the trans-medical meaning of medicine – that is, with a theory of medicine and its fundamental axioms, particularly those which underlie the clinical experience. Medicine cannot examine its own meaning and cannot address itself as a "problem." What values and axioms are at the root of medical thought and action? What is the meaning of health and disease? These are not meaningless questions, since they are so much a part of every human existence and the substrate for so much human action and attention. What idea of man undergirds medical theory and practice, and what idea of man emerges from medical observations? What is the physician and what is his social function?

There are questions in abundance, and there is room for many modes of philosophic inquiry, from the analytic to the phenomenological, as well as the formal and speculative. Such inquiries will benefit philosophers, as well as medicine and society. New fields of human experience will be opened to philosophic inquiry, of course. But above all, philosophy will be enriched by its encounter with the vast body of factual and particular knowledge from a large variety of viewpoints which medicine now possesses about man and his existence.

Out of the engagement of philosophy with medicine, one special branch of philosophy can be energized in particular. This is the field of philosophical anthropology which has had somewhat dubious origins in the Enlightenment's science of human nature and twentieth century German philosophy. The pretensions of philosophical anthropology need to be trimmed back, but if this can be accomplished, there might emerge a new synthesis of the image and idea of man suitable to our times. Medicine stands ready to provide grist in abundance for the philosophic mill. Philosophy, now cleansed of its speculative excesses, provides a more efficient and surer mill for the new grist.

To stretch the point just a bit – and consciously to titillate your fancy, we might even need medicine to help us understand philosophy! Miguel de Unamuno may have overstated the case when he said, "Philosophy is a product of the humanity of each philosopher.... And, let him do what he will, he philosophizes not with the reason only, but with the will, with the feelings, with the flesh and with the bones, with the whole soul and the whole body. It is the man that philosophizes"[6].

If Unamuno is even partly right then, philosophy itself may require some explanation in terms of health or disease, or at least gain insights from perceptions which are the special domain of medicine. Someday it may not be whimsical to speak of a medicine *of* philosophy, as well as a philosophy *of* medicine.

I will stretch your credulity and patience no further. Clearly, I see philosophy as important to medicine as medicine is to philosophy. Philosophy *in* and *of* medicine offers opportunities for both disciplines to mature and to contribute to our understanding of man and human culture. Indeed, without the engagement and the conjunction of medicine and philosophy, no viable or understandable image of man can be synthesized for our times. And, the absence of such a synthesis is a major deficit in contemporary culture.

NOTES

[1] Robert Burton, *The Anatomy of Melancholy* (New York: Tudor, 1955), p. 279.

[2] Edmund D. Pellegrino, "Medicine and Philosophy: Some Notes on the Flirtations of Minerva and Aesculapius," Annual Oration of the Society for Health and Human Values (Delivered: November 8, 1973, Washington, D.C.).

[3] José Ortega y Gasset, *The Mission of the University* (New York: William Norton and Co., 1966).

[4] Scott Buchanan, *The Doctrine of Signatures – A Defence of Theory in Medicine* (London: Kegan, Paul, Trench, Trubner, 1938).

[5] Michel Foucault, *The Birth of the Clinic, An Archeology of Medical Perception*, (New York: Pantheon Books, 1973).

[6] Miguel de Unamuno, *The Tragic Sense of Life* (London: The Fontana Library, Macmillan, 1962), p. 45.

NOTES ON CONTRIBUTORS

Chester R. Burns, M.D., Ph.D., is James Wade Rockwell Assistant Professor of the History of Medicine, Institute for the Medical Humanities, and Assistant Professor in the Department of Preventive Medicine and Community Health, The University of Texas Medical Branch, Galveston, Texas.

H. Tristram Engelhardt, Jr., Ph.D., M.D., is Assistant Professor of the Philosophy of Medicine, Institute for the Medical Humanities, and Assistant Professor in the Department of Preventive Medicine and Community Health, The University of Texas Medical Branch, Galveston, Texas.

Samuel Gorovitz, Ph.D., is Professor of Philosophy and Chairman of the Department of Philosophy in the University of Maryland, College Park, Maryland.

Patrick A. Heelan, Ph.D., is Professor of Philosophy and Chairman of the Department of Philosophy in the State University of New York at Stony Brook, Stony Brook, New York.

Lester S. King, M.D., formerly Senior Editor of the *Journal of the American Medical Association*, is now Contributing Editor; he is Professorial Lecturer in the History of Medicine at the University of Chicago.

Loretta Kopelman, Ph.D., is an Instructor in the Department of Pediatrics at the University of Rochester School of Medicine, Rochester, New York.

Alasdair MacIntyre, M.A. (Oxon.), is University Professor of Philosophy and Political Science, Boston University, Boston, Massachusetts.

Edmund D. Pellegrino, M.D., is Chancellor of the University of Tennessee Medical Units, Professor of Medicine and Humanities in Medicine, and Vice-President for Health Affairs for the University of Tennessee System, Memphis, Tennessee.

André Schuwer, Ph.D., is Professor of Philosophy and Chairman of the Department of Philosophy in Duquesne University, Pittsburgh, Pennsylvania.

H. T. Engelhardt, Jr. and S. F. Spicker (eds.), Evaluation and Explanation in the Biomedical Sciences, 235–236.
All Rights Reserved. Copyright © 1975 by D. Reidel Publishing Company, Dordrecht-Holland.

Jerome Shaffer, Ph.D., is Professor of Philosophy, The University of Connecticut, Storrs, Connecticut.

Stuart F. Spicker, Ph.D., is Associate Professor of Philosophy in the Department of Community Medicine and Health Care, School of Medicine, University of Connecticut Health Center, Farmington, Connecticut.

Stephen Toulmin, Ph.D., is Professor of Social Thought, The University of Chicago, Chicago, Illinois.

Kenneth Vaux, Dr. Theo., is Professor of Ethics and Acting Director of the Institute of Religion and Human Development at the Texas Medical Center in Houston, Texas.

Marx W. Wartofsky, Ph.D., is Professor of Philosophy, Boston University, Boston, Massachusetts.

Richard M. Zaner, Ph.D., is Easterwood Professor of Philosophy and Chairman of the Department of Philosophy of Southern Methodist University, Dallas, Texas, and Adjunct Professor of the Philosophy of Medicine in The Institute for the Medical Humanities, University of Texas Medical Branch, Galveston, Texas.

INDEX